POLITICS ON THE ENDLESS FRONTIER

POLITICS ON THE ENDLESS FRONTIER

Postwar Research Policy in the United States

DANIEL LEE KLEINMAN

Duke University Press Durham and London 1995

Printed in the United States of America on acid-free paper ∞

Designed by Cherie H. Westmoreland

Typeset in Plantin Lt. with Futura Book display

by Keystone Typesetting, Inc.

Library of Congress Cataloging-in-Publication Data appear

on the last printed page of this book.

A small portion of chapter one appeared in

Social Studies of Science (1991, vol. 21) and

portions of chapters three, four, and five appeared in

a very different form in *Science, Technology,*

and Human Values (1994, vol. 19).

For Flora, who added wonder and

For Susan, who added garlic

CONTENTS

TABLES

ACKNOWLEDGMENTS

Many people contributed to this book. That it is completed, coherent, and, I hope, cogent is a tribute to Jack Kloppenburg. Since we met in 1985, Jack has provided unwavering support, contagious enthusiasm, and thoughtful advice. He is an extraordinary teacher, an insightful and spirited colleague, and a steadfast friend.

My conversations with Charles Camic have made the arguments in this book stronger and the writing smoother. Marc Schneiberg spent many hours helping me work out the book's narrative logic. Larry Cohen and Susan Bernstein listened and read and provided important emotional support. The institutional orientation I take owes a great deal to graduate courses and discussions I have had with Leon Lindberg. When I was looking for an academic space in which to explore the issues that concerned me, Mark Gould introduced me to sociology.

Several people have read and provided comments on portions of the book. They include Lisa Brush, Warren Hagstrom, Allen Hunter, Lane Kenworthy, Ann Orloff, Mark Solovey, and David vanKeuren.

I completed most of the revisions on the book during a year spent at the University of Houston—Clear Lake. This time was made enjoyable and productive in large measure due to the support of three colleagues—Barbara Butler, Pamela Garner, and Douglas Holmes.

In addition to the support of friends and colleagues, I could not have completed this project without the assistance and support of library personnel. Staff at the Hagley Museum and Library, the Library of Congress, the Franklin Delano Roosevelt Library, the Wisconsin State Historical Society Library, and the University of Wisconsin–Madison libraries were all very helpful. I benefited especially from the expertise of Michael Nash at the Hagley.

At Duke University Press, Lawrence Malley first took an interest in my project. Reynolds Smith has shepherded the manuscript through the editorial process. The press's two anonymous reviewers provided careful and useful critiques of the entire manuscript, and I believe the work is much stronger as a result.

Former Vannevar Bush aides John Connor and Lloyd Cutler graciously agreed to be interviewed on their involvement in the establishment of the National Science Foundation. Several policymakers and "opinion leaders" in Washington also agreed to be interviewed on recent research policy debates. Those interviewed include Daniel Burton, William Cunningham, William Krist, Denise Michel, Pierre Perrolle, Mark Schaefer, William Stiles, Al Teich, James Turner, and Leonard Weiss.

The process of doing the research and writing the initial manuscript was made easier by financial support from the University of Wisconsin–Madison, the National Science Foundation, and the Hagley Museum and Library.

In the very final stages of preparing this manuscript, I benefited from the assistance of Kathy Michaelis and the financial support of the School of History, Technology, and Society at the Georgia Institute of Technology.

POLITICS ON THE ENDLESS FRONTIER

1

THINKING ABOUT THE POLITICS OF SCIENCE AND SCIENCE POLICY

> Science knows no politics. Are we in this frenzy of economy, brought about by those who control the wealth of the country, seeking to put a barrier on science and research . . . ? Science will go on when existing political parties will long have been forgotten. . . . Do not seek to put the hand of politics on these scientific men who are doing great work.
>
> Fiorello La Guardia, New York Mayor (1934–41)

> The "pure" universe of even the "purest" science is a social field like any other, with its distribution of power and its monopolies, its struggles and strategies, interests and profits. . . .
>
> Pierre Bourdieu, 1975

In the early years after World War II, the results of scientific research were widely viewed as a means to social and economic salvation—a way to improve national welfare and individual lives. Agricultural production increased dramatically as the result of research in agrichemicals and plant breeding. The development of new drugs and medical techniques led to the control and elimination of many diseases. Nuclear power promised an inexpensive solution to burgeoning postwar energy needs. Chemical companies promised "better living through chemistry." In this context, the idea that there was a politics of science[1] must have seemed absurd.

The successes of scientific research appeared almost magical after World War II. The activities of scientists seemed to exist somehow outside of society—a realm different from any other, not shaped by power and bias, but by the search of all knowing experts for truth and the means to social progress (Jones 1975). Little public attention was given to who controlled science. The "results" of research were enough to indicate to most people that science and scientists were beneficient. This context was well suited to the belief widely held by scientists that scientific practice was and should be autonomous from the rest of society and that the scientific community ought to be accountable only to itself (Merton 1973; Polanyi 1951, 1962).

Since the optimistic postwar years, unambiguously negative environmental, social, and economic impacts of scientific research and technology[2] have led to public misgivings about science and to challenges to scientists' autonomy. In areas as wide ranging as nuclear power, biotechnology, and AIDS research, activists and citizens have repeatedly questioned the right of experts to make decisions affecting their lives. And big science—capital-intensive research—is faced with a fiscal ceiling of its own. In a time of economic crisis and belt tightening, government officials and concerned citizens wonder if massive investments in capital for scientific research (for example, the superconducting supercollider and the space station) will pay off.

Leaders in the scientific community recognize the threat to their resources and autonomy that is posed by public dissatisfaction and government fiscal crises. After World War II, in a report entitled *Science—The Endless Frontier* (1945), Vannevar Bush and his elite scientist colleagues promised economic and social prosperity in exchange for generous government support for scientific research and effective scientist control of allocation of these resources. Nearly half a century later, in his own report, the outgoing president of the American Association for the Advancement of Science (AAAS), Leon Lederman, wonders if science is at "the end of the frontier" (1991). Like Bush, he suggests that national prosperity depends on high levels of government support for scientific research and scientist autonomy. He argues that, while the resources science needs to make good on its promise of national welfare through research are growing, availability of resources is not keeping up.[3]

At a moment when science is coming under increasing scrutiny, the American economy is faced with a so-called crisis of competitiveness. The causes of this economic crisis are multiple. Internationally, economic growth is slower than it was in years past (Greenhouse 1987: 1). In addition, U.S. manufacturing capacity has eroded (Cohen and Zysman 1988). Some analysts view the federal budget deficit, which absorbs potential investment capital, as a central problem (Baumol 1989). Other policy experts see the primary source of the problem as organizational—the failure to fit production organization to current economic realities (Cohen and Zysman 1988). Still other opinion leaders argue that the pace of technological development is itself the source of the problem (*Business Week* 1982: 126–30). Indeed, a 1990 study by the Department of Commerce found that while the United States was competitive with Japan in terms of research and development on twelve emerging technologies, in the introduction of these technologies for production, Japan was ahead in several areas and is pulling ahead in others (*Science* 1990: 1185).[4]

Leading analysts of our current economic situation, Michael Piore and Charles Sabel (1984) argue that we are at a *second industrial divide*. Mass production has brought us vast national wealth, but we have reached its limits. We need a new production system—a system of permanent innovation—based not just on new organization, but on new technology and new policy arrangements.

In the face of economic crisis and national soul searching about science, science policy and the existing mechanisms for making policy have been sharply criticized. In the context of fiscal restraint and economic malaise, there has been a good deal of debate over the need to coordinate and plan science and technology policy and to set priorities for federal support of basic and technology research. In 1988, a congressional task force concluded that "we lack adequate long-range coordinated government policies directed toward future growth and prosperity" (U.S. House Technology Policy Task Force 1988: 4). The congressional body called for the establishment of a new government organization capable of "prioritizing new advanced technology initiatives and technology" on the basis of guidance from business and labor (12). In 1991, echoing the task force's concerns, a coalition of high technology trade groups and businesses called on the administration to work with it "to establish formal mechanisms for private sector involvement in setting federal R&D priorities" to maximize the benefits of federal R&D investment.[5] And President Clinton is moving in this general direction.

World War II led to a widespread consensus that basic research was the foundation for economic prosperity and thus merited government support. While this view was widely held among business elites after the war, high-technology industry is today focusing on government support for research on critical technologies, not basic research. And, while some scientists hark back to this early postwar perspective (e.g., Lederman 1991), not all are so utopian as to expect increased funding with no strings attached. In the face of dwindling federal resources, some scientists are now setting their own priorities by field of study to maintain their autonomy and head off external meddling in priority setting.

But there is not even consensus on whether scientists should define disciplinary or community-wide research agendas. A 1991 study by the Congressional Office of Technology Assessment (OTA) argues that scientists may not be the best group to set priorities for research. According to the report, scientists lack the appropriate organizational arrangements for evaluating projects across disciplines. Moreover, the OTA argues that scientists may not be the best people to evaluate such issues as project timeliness and the social and economic benefits of diverse projects (Office of Technology Assessment 1991). The report suggests that Congress should hold biennial hearings on the state of the research system to determine when the federal research "portfolio" is out of balance.

In a 1993 report, the National Academy of Sciences proposed its own framework for establishing national research priorities (National Academy of Sciences et al. 1993). The report acknowledges that elected officials should determine which fields of research are most important, but it proposed that decisions on research funding should be framed by concern for U.S. leadership in *all* major areas of scientific research and recommended that panels of experts determine the relative strengths and weaknesses of individual fields.

While not always addressing the issue of priority setting specifically, some activists support increasing grass-roots decision making in science policy. According to scholar and activist Dorothy Nelkin, "democratic ideals imply that science policy be subject to greater public scrutiny and political control" (1984: 36). Such calls for increased democratic participation are heard in areas as wide ranging as the determination of AIDS research protocols and university research agenda setting.

A good deal of social science research has shown that "crises" are opportunities for institutional change. Normally, organizational structures or institutions are stable, changing only incrementally (Hall 1986: 266). But during crises—such as wars, economic depression, or widespread social strife—institutional innovations or fundamental changes and reorganization are possible. Indeed, changes in the character, if not organization, of science and technology policy are now under way in the United States.

The notion of science policy is fairly broad. It may include the establishment of research priorities for government-supported research, as well as the regulation of the potentially negative effects of technological development (BLR Smith 1990: 6). In this book, I focus on the former, which I term *research policy,* because I believe that how and what research priorities are set has a dramatic impact on the contours of a society and, indeed, says a good deal about the society's broader values and priorities.

New York City Mayor Fiorello La Guardia, in the quotation that led off this chapter, insisted that "science knows no politics." To the contrary, I believe that science knows a very definite, if historically variable, politics. In this book, I explore the nature of that politics in a particular context: the rise of a science elite and the genesis of the federal research policymaking system after World War II. I probe the values that characterized the collective advancement project of scientists after the war and explain how they became embodied in the postwar research policymaking establishment. I investigate the factors that led to the postwar federal research policymaking establishment and in so doing, theorize and empirically characterize important components of the scientific field in the United States from about 1900 to 1950.

But the analysis does not stop in 1950. It includes consideration of present-day struggles to reorganize the postwar federal research policymaking establishment. I assess the degree to which the likely outcomes of these struggles are constrained by institutional and discursive legacies dating back half a century, as well as the degree to which changed conditions may alter the outcome of this struggle.

Back to the Future

Recent debates over the organization of research priorities—especially the links made between national economic welfare and scientific research—

are reminiscent of debates that raged from the early 1940s through 1950. In all spheres, we tend to view our current situation as the historically necessary or inevitable product of the past. But the present is the product of crucial choices and decisions made at critical points in the past (Ikenberry 1988: 224; Kloppenburg 1988; Noble 1984; Piore and Sable 1984). These choices, although institutionally constrained, are not pre-given. And as we look forward to the reorganization of research priorities in the United States, it makes sense to look back to the postwar debates that defined the research policy establishment which sets federal research priorities today.

The argument made today by business interests and politicians—that the nation's economic welfare might be improved by a *centralized* research policy organization that would make possible coordination and planning—is not new. Indeed, this was a common theme as the prospects for a new research policymaking institution were debated in Congress and in the scientific community for nearly a decade beginning just prior to U.S. involvement in World War II. Popular concerns about industrial monopoly and about the need to harness technological innovation to the objectives of economic growth and reform prompted calls for a single *centralized* and politically responsive organization to *direct* and consciously *plan* the advance of scientific research and technology. Many scientists and representatives of large-scale business, however, favored a more decentralized and laissez-faire system.

Beginning in 1942, Senator Harley Kilgore (Democrat, West Virginia) made numerous legislative proposals for the development of a single centralized organizing mechanism—one leading federal agency—that would *coordinate* research in an effort to promote economic development. Kilgore wanted this organization to be *democratically controlled* and suggested that it be administered not exclusively by scientists, but by representatives of a wide range of social interests. He argued that patents resulting from government-sponsored research ought to be the property of the federal government, which would have the sole right to license them.

Kilgore had allies, especially among New Deal liberals and significant portions of the Roosevelt and Truman administrations. But he also faced opposition from elites in the scientific community, military leaders, political conservatives, and certain segments of the business community. This opposition, led by Vannevar Bush, wartime head of the Office of Scientific Research and Development (OSRD) and one-time MIT vice-president, favored an organization controlled by scientists, which would promote basic research, not necessarily economically relevant research. In the end, Kilgore and his allies were largely defeated in the struggle. The creation of the National Science Foundation (NSF) in 1950 and the maintenance of a system of disconnected mission-oriented agencies was the outcome of the conflict.

At the outset of the debate over creation of a new research policymaking system, the notion of a "system" really meant a single agency. Kilgore, in particular, but to a lesser extent Bush and his allies as well, had in mind a single *centralized* agency. By providing an account of the debate over this single agency I am really providing an analysis of the broader system. As I will show, it was, to a considerable extent, the delay in passing legislation to create a single agency—variously termed the Office of Scientific and Technical Mobilization, the National Research Foundation, and the National Science Foundation—that led to a fragmented "system" of research policymaking composed of several government agencies, new and existing.[6] There was never a debate over a research policymaking system as such. There were several debates—one over expanding the Public Health Service, another over creation of the Atomic Energy Commission, and so forth. But the controversy over the creation of what came to be termed the National Science Foundation was pivotal in shaping the contours of the postwar federal research policymaking system overall.

In the course of providing an analysis of the genesis of the federal postwar research policymaking system in the United States, I refer interchangeably to the conflict over the creation of a research policy agency, creation of a research policy system, and creation of the National Science Foundation. In all three cases, I am referring to the same debate—the controversy over what came to be called the National Science Foundation.

Intellectual Context and Methods

In carrying out this project, explicitly and implicitly I draw on and react to several distinct, though related, literatures. I consider this study a political sociology of science policy. Hence, my analysis is informed by recent developments in both political sociology and the sociology of science. In addition, I rely heavily on work in the history of science policy. I have learned a great deal from these literatures, and I utilize and extend them in providing my explanation for the establishment of the configuration of federal agencies designated and/or created to make research policy in the period after World War II.

Below, I briefly summarize the essential elements of each line of thought. I then provide an explanation for the genesis of the postwar configuration of federal research policymaking institutions that follows from each approach. I conclude my discussion of each of the literatures with consideration of what I believe is missing from the type of explanation it provides. Finally, in the section that follows I provide an outline of my own explanation for the phenomena under consideration. This framework is elaborated in detail in the chapters that follow.

Science and Society

Like the general public in the immediate postwar period, early sociologists of science viewed science as fundamentally distinct from other social realms. Arguing that science was unlike the rest of society where decisions were made on the basis of the ascribed characteristics of actors and the non-merit-based power some actors have over others, Robert Merton (1973) and his colleagues asserted that the scientific community was governed by universalistic norms and was normally meritocratic. Given the distinctiveness of science, it logically followed that only scientists could make decisions about science. Public intervention generally was viewed as inappropriate (Polanyi 1951: 53; 1962: 67, 72).

Recent work in the sociology of science has successfully challenged the Mertonian contention that science is fundamentally distinct from other social realms. The idea of a completely insulated meritocratic realm has been replaced by the idea that "science is a field like any other" (Bourdieu 1975: 19), and as Bourdieu suggests, it has "its distribution of power and its monopolies, its struggles and strategies, interests and profits" (1975: 19).[7]

For many, challenging the distinctive character of science has meant showing that the so-called natural world is itself socially constructed and that knowledge does not unambiguously reflect some external reality, but is the product of social processes (Barnes 1974; Bloor 1976). From pointing out the constructed character of scientific knowledge, research turned to understanding the social processes through which what "we come to call knowledge is constituted and accepted" (Knorr-Cetina 1983: 116; Latour and Woolgar 1979).

This "constructivist" work in the sociology of science marks a considerable advance over earlier work, but it suffers from several important and related weaknesses. In general, concern with knowledge or fact construction has led analysts to the level of the laboratory, and the focus is on microlevel—laboratory—politics. Typically, little attention has been paid to the linkages between the level of the laboratory and the economy, as well as that of the state. Work in the history of science and technology has clearly shown that understanding this more macrolevel is essential to understanding how research agendas are shaped and therefore how facts and artifacts are constructed (Foreman 1987; Kloppenburg 1988; Kohler 1979; Leslie 1993; Noble 1984).

Some effort has been made to link the macro and the micro. Latour, whose book with Steve Woolgar on fact construction focused almost exclusively on the laboratory level (Latour and Woolgar 1979), has described successful scientific achievements as the product of enrolling actors—human and nonhuman, living and nonliving—to one's cause (1987). Success in science, according to Latour, depends on the development of powerful alliances or networks that extend beyond the laboratory. These networks include not only interested human actors, but what Latour refers to as nonhuman actors—microbes, electrons, diseases, animals, and the like. But Latour never indicates where the capacity to enroll actors to one's cause comes from. In general, he never addresses the reasons why some alliances might work and others might not in a struggle over fact construction. More specifically, Latour does not consider how alliances are structured by power relations between actors and what role these relations play in determining success in struggles over fact construction. As Amsterdamska has noted, "without making distinctions among the kinds of control to which the different categories of actors and actants can be subjected, and among the various means by which domination can be exercised over different kinds of actors and actants," we cannot possibly understand why certain actors are triumphant in science, while others fail (1990: 501).

Constructivist approaches and laboratory studies have dominated the sociology of science for the better part of a decade. However, interest has increased in linking laboratories with other social arenas (Cozzens and Gieryn 1990). Still, this work has not typically focused on explaining science policy outcomes and state building. There are, of course, exceptions. Cambrosio, Limoges, and Pronovost (1990) have attempted to utilize the approach of Latour and others in a move away from the level of the lab to investigate the construction of science policy.[8] Cambrosio and his colleagues seek to explain the establishment of the biotechnology dossier as a matter of Quebec government science policy. Theirs is really a study of category construction, a study of the process of classification. The authors trace the process from the drafting of an action plan through its ultimate approval by the provincial cabinet. The process leading to ultimate approval is never entirely clear in Cambrosio et al., but there are hints at explanation. The authors suggest that categories are constructed "by enrolling external actors in a collective endeavor" (1990: 213). Thus, the winners in a struggle over category construction—and policy construction—are those best able to enroll allies, human and nonhuman, who support their construction. Another part of the explanation for victory in policy construction appears to be the creation of an intertextual web which, by connecting a set of documents, may provide the basis for the new document's legitimacy.[9]

How might such a framework be deployed in explaining the struggle over the creation of a research policymaking apparatus in the United States after World War II? I believe we would begin with two texts: the original proposal made by Senator Harley Kilgore in 1942 and the alternative proposed by Vannevar Bush and his colleagues in 1945. The fact that the final legislation—the National Science Foundation Act of 1950—was largely a mirror of the Bush-inspired proposal would suggest that Bush and his colleagues were better able than Kilgore and his supporters to enroll allies to their cause. Certainly Bush had a broad range of allies in the state, in industry, and among scientists. The Bush cause may also have benefited from some form of intertextual web that strengthened the philosophical underpinnings of the victorious legislation. The one major compromise made to Kilgore and his supporters in the final legislation—giving the president the right to appoint the NSF director—could be seen as a way to weaken Kilgore's alliance and move the senator's human allies and, indeed, the senator himself, over to the Bush camp. Victory is the product of a broad and powerful network of allies, human and textual.

As the work of Cambrosio and his colleagues suggests, I believe we must consider the role of discourse in shaping the outcome of any legislative dispute. Although Cambrosio et al. assert the importance of drawing on texts with authority, their explanation would be strengthened if they indicated the basis for the authority of the texts on which they focus. In the case under study here, Bush and his colleagues drew on the broad social authority of science in constructing their legislative proposals. As I will show, this authority was the product in part of the military successes of science during World War II and of a more general belief in the power of science. Textual authority does not come out of thin air, but has social and historical bases.

The focus on ally enrollment within the framework provided by Cambrosio and his colleagues does not—indeed, cannot—allow us to understand why it was possible for Bush to enroll actors and not Kilgore. The approach does little to help us understand why, if victory depended so considerably on enrollment, it took eight years to achieve a victorious alliance. Alliances are clearly central to understanding the shaping of postwar research policymaking in the United States. But, as I will show, their character is shaped in important ways by the relationship between civil society and the American state. And this in turn depends considerably on the structure of each. Finally, understanding the nature of alliances depends on understanding the interests or projects of actors, and Cambrosio and his colleagues pay little attention to these.[10]

History of Science and Government

There is an extensive literature in the history of science area that treats the relationship between science and the federal government. And we need not simply speculate on how researchers in the area might explain the genesis of the federal postwar research policymaking complex or more specifically what came to be the National Science Foundation (NSF), since there are many books, articles, and dissertations that explore the founding of the NSF (England 1982; Kevles 1977, 1987; Maddox 1981; McCune 1971; Pursell 1976, 1979a, 1979b; Rowan 1985).

Historians who do work on government and science tend to focus primarily on important individuals. The work that most directly addresses the struggle to establish a National Science Foundation portrays the struggle as a political battle largely between two men, Harley Kilgore and Vannevar Bush, and ultimately attributes the defeat of Kilgore's proposals to a single congressional vote. Some of the best work on the controversy over the establishment of the NSF recognizes that actors' interests are defined by social position (Kevles 1977, 1987), but I believe this work can be enriched by adding consideration of the role various political alliances, relations of power, and organizational constraints played in determining the legislative outcome.

My work owes a great deal to research in the history of U.S. science policy (England 1982; Kevles 1977, 1987; Maddox 1981; McCune 1971; Pursell 1971, 1976, 1979b; Rowan 1985; Schaffter 1969). I build on this work, considering the *social structural environments* in which actors act, and I consider how the histories of these structures shape interests and define possibilities. For example, while the existing literature does not give much attention to U.S. party politics, I explore how the U.S. party system with a history rooted in nineteenth-century patronage politics (Shefter 1977; Skowronek 1982) shaped the outcome of the struggle between two competing philosophies (embodied in Bush and Kilgore) for the organization of science policy.

The existing historical literature forms an important starting point for my work, but, as Ikenberry notes:

> It is not enough simply to explore the immediate struggle over policy by societal, governmental, and transnational actors. While the process of policymaking is important, that process itself rests upon larger structures that influence, guide, redirect, magnify, and inhibit policy battles. Consequently, we need a better appreciation of the shaping and constraining role of the policy or institutional setting, as well as the historical dynamics that shape this institutional setting. (1988: 222)

Following Ikenberry's comment, I explore the institutional environment in which the struggle to establish the National Science Foundation occurred.

State and Society

The sociological debate on the nature of the state in recent years has been dominated by researchers working in the so-called state-centered tradition. Scholars taking this approach have forcefully challenged the hegemony of neo-Marxist analyses of the state,[11] and they have rejected the implication of much of this work that the capitalist class is omniscient and omnipotent and that the state will always function to reproduce capitalist relations (see Skocpol 1980). In addition, Skocpol and others have criticized this work on the state for paying insufficient attention "to variations in state structures and activities across nations and . . . time periods" (1985: 5; see also Campbell 1988: 11).[12]

In the case of the legislative battle to establish a single central agency to make federal research policy in the postwar period, Skocpol and her colleagues might point out first that more was at stake in this legislative battle than capital accumulation.[13] At the very least, we must acknowledge the aim of elite scientists to gain autonomous control of federal resources for research.[14] Beyond the issue of reducing legislative struggles to struggles over capital accumulation, one variant of a state-centered approach would stress the importance of state and party structure in explaining the outcome of this policy dispute. Such an analysis might highlight the importance that lack of party discipline played in delaying passage of legislation for some eight years. Kilgore was a Democrat; both houses had Democrats in the majority for most of the period, and Kilgore had the support of president Truman. But Democrats were divided, and conservative Democrats often helped Republicans block Kilgore-inspired measures.

Advocates of a state-centered approach would also stress the specific nature of the U.S. state—its fragmented and highly permeable character. Whereas many European parliamentary systems promote unified voice between the legislative and executive, the structure of the U.S. state makes possible division between the Congress and the executive and within each. Such divisions were quite visible during the debate over NSF legislation. What is more, the permeability of the American state—the multiple points (congressional committees, federal agencies, and so on) at which actors can make their interests heard—made it easy for elite scientists and their conservative congressional allies, not constrained by disciplined political parties, to stop measures they found unacceptable.

The eight-year delay in passage of legislation to establish a postwar research policy agency in the United States—what came to be called the National Science Foundation—can be understood to a considerable degree in terms highlighted by sociologists often associated with the state-centered approach. The permeable and fragmented character of the U.S. state, combined with undisciplined political parties, led to one delay after another in passage of research policy legislation. Two bills died in committee at the hands of cross-party coalitions, and one bill was vetoed by a Democratic president presented with a Republican piece of legislation. When the final bill was signed eight years after Kilgore first proposed a single comprehensive science agency, many of the responsibilities Kilgore had envisioned for his agency had been taken up by agencies created in the meantime with narrower, highly circumscribed, responsibilities.

I start with state and party structure as state-centered theorists recommend and add to this consideration of the *formal and informal* aspects of U.S. policymaking and how they are related to the particular structure of the American state.[15] I focus attention on the direct and indirect roles of elite scientists and representatives of industry—their roles inside and outside the state—on the ultimate policy outcome.[16]

As I will show in the chapters to come, in order to understand state building and policymaking in the United States, we must consider how the structure of the state shapes and refracts the influence of social groups. Given nonprogrammatic political parties and a fragmented and highly permeable state, business and other social interests—here, scientists—are more likely to draw on informal contacts to shape policy than on formal organizational relations with the state. In addition, where the state structure and the line between it and civil society is highly permeable, the formal role of social interests in influencing policy may come when they obtain positions as state managers rather than through, for example, direct contacts between the state and peak associations.

Policymaking and Informal Networks

Skocpol's critique of structuralist Marxists makes sense, and she has provided a quite admirable empirical assessment of the basic tenets of structuralist Marxism (1980). But before the structuralists came on the scene, researchers in the corporate liberalism tradition—what disparagingly came to be termed "instrumentalism"—dominated the area of critical studies of the state. Although I find this early tradition wanting in certain respects, I believe its easy dismissal is inappropriate.[17]

The argument championed in various forms by William Domhoff (1979), Ralph Miliband (1969), and others is essentially that capitalists can often shape policy by their *direct participation in the state*—in the policymaking process. This contention suggests the need for close empirical research that focuses on the personal connections among elites and their role in the state.[18]

In the present case, following scholars like Domhoff and Miliband, we would seek to understand the final outcome in the debate over research policy legislation—the National Science Foundation—as the product of elite maneuvering inside and outside the state. And indeed, this approach explains a good deal. There are strong connections between Bush and his colleagues and corporate leaders, and these contacts surely shaped Bush's policy agenda. Evidence suggests elite scientists inside and outside the state played an important role in stalling legislation, and there is some evidence of business involvement in directly shaping Bush's legislative program.

This approach provides us with important insights. However, simply showing social connections does not prove that they had a meaningful effect on the outcome, and this is often where advocates of this approach leave off. In addition, elite scientists were not simply the tools of the capitalist class as a simple Marxist analysis might conclude. Rather, their social location was in part independent of the location of the capitalist class, and additionally this group had autonomous objectives—objectives that cannot be reduced to their relationship to the capitalist class. Finally, a more thoroughgoing account must understand where and why personal connections mattered in determining the policy outcome.

The Structure of the Analysis

In the chapters that follow I engage in synthesis and extension of the literatures I have just discussed in order to explain the genesis of a postwar research policymaking apparatus in the United States. The account I provide should be viewed simultaneously as a study of state building in the United States and as a study of an effort by elite scientists to realize their project to extend scientist control over resources for government-supported research.

My explanation suggests that four general factors shaped the way research policymaking came to be organized in the United States in the period after World War II. These factors are the configuration of the prewar scientific field,[19] World War II itself, the collective advancement project of U.S. scientists, and the structures of the U.S. state and civil society and the relationship between the two.

My approach to the problem of the genesis of a postwar research policymaking agency in the United States puts power at its core. But power has a source; it comes from somewhere. In the analysis that follows, I argue that power must be understood in an institutional context. The institutional configuration of a society, including the structure of the state, affects the degree of power any set of actors can have in determining policy outcomes (Hall 1986: 19, 231). Leon Lindberg provides a useful definition of institutions:

> Institutions are not simply aggregates of individual preferences or passive mechanisms for transmission of economic impulses, but constitute a historically specific constraint and opportunity structure that implies an enduring division of labor and rules of play, that establishes distinctive capacities and incapacities, and that constrains the strategies any individual, economic agent, or political authority can adopt to achieve its aims. (1982: 24)

Thus, the manner in which the state and social actors are organized will provide opportunities and constraints, and it is in terms of these opportunities and constraints that actors' power should be understood (see also Ikenberry 1988: 223). In addition to defining the capacities and incapacities of actors, institutional location plays an important role in defining the objectives of actors' collective advancement projects. Groups—be they state managers, classes, class fractions, or professions—are bound by the advancement goals they share as a product of shared institutional location.

An investigation of the genesis of a federal postwar research policy agency—ultimately named the National Science Foundation—must begin with the prewar scientific field. I view the field as a shifting configuration constituted by the overlap and interaction of several institutional spheres: foundations, universities, science-based industry, and the state (Kleinman 1991). In some sense, these spheres can be viewed as distinct, with independent logics and actors with specific collective projects.

The prewar scientific field in the United States affected the (re)organization of research policymaking in the postwar period in at least four ways. First, it shaped the collective advancement project of elite scientists and the interests of science-based industry. Second, it provided institutional legacies and policy histories on which actors drew in specifying their proposals for, and demands of, a postwar research policymaking agency. Third, the prewar scientific field provided the basis for the development of a range of social connections—social capital—on which actors drew in their efforts to shape the postwar field. Finally, the prewar field provided a context within which scientists could develop the credibility—scientific authority—that facilitated their entree into government during World War II.[20]

Prior to World War II, the group of elite scientists—largely science administrators who ended up controlling wartime research policy and played a central role in shaping the contours of postwar research policy—became prominent through achievements in bench research and science administration. These men were at the center of the prewar scientific field: they became the administrative leaders of the nation's most important research universities and research institutes; they were voted positions on the boards of prominent science-based firms; they were selected to advise the federal executive on science issues; and they served and were elected to membership in such prestigious organizations as the National Academy of Sciences.[21]

The prewar connections with the federal executive put them in a position to gain control of wartime research policymaking and ultimately to shape postwar research policymaking. This group was in close contact with industry research leaders, and the patent policy they advocated after the war was fully in line with the position taken by the science-based firms with which they had intimate contact prior to the war. Certainly their views on this issue were shaped by their relationship with industry leaders.

Prior to the war, several U.S. industries developed autonomous research capacity. This defined their interest in a postwar policy that maintained government support for basic research, but not applied. Firms from these industries focused on applied research. They wanted no competition from the government in this area, but saw a useful role for federal support of basic research—research that was for them not directly profitable, but ultimately an important basis for their own work. Firms from industries that developed research capacity prior to World War II were at the forefront of business efforts to restrict postwar government funding to basic research.

In the proposals of both Bush and Kilgore, institutional legacies and policy histories were clearly important. The big foundations, Rockefeller and Carnegie, served as a model for postwar research policymaking. The centrality of scientists in World War I research policymaking established the legitimacy of scientist control of research policy—an objective Bush and his colleagues achieved in the final NSF legislation. In addition, World War I science agencies legitimized the concentration of research support in a few research universities and established what would later be an important principle of research by contract. Contract research allowed scientists to remain at their "home" universities rather than work in some central government laboratory. Finally, the importance of central coordination of research policymaking was established as a result of precedents set during the first World War. *Central coordination* became a pivotal framing issue in the debate over the organization of postwar research policy.

If the period prior to World War II defined the interests and projects of scientists and science-based industry, established a set of institutional and policy legacies that shaped debate over a postwar research policy agency, and defined an initial set of social connections, the war itself was a crucial turning point in the effort to define an agenda for postwar research policy. Organizational structures or institutions are generally stable and most often only open to incremental change (Hall 1986: 266). It is during periods broadly recognized as *crises* that important institutional innovations or fundamental changes and reorganization are possible. As Skowronek notes: "Crisis situations tend to become watersheds in a state's institutional development. Actions taken to meet the challenge often lead to the establishment of new institutional forms, powers and precedents" (1982: 10). During crisis periods, struggles over the basic rules of the game become possible (Krasner 1984: 234).

It is within the context of crisis, then—the prospect of U.S. entry into World War II—that the emergence of the postwar research policy establishment and the related project of America's elite scientists must be understood. During the war, Vannevar Bush and his colleagues parlayed the social capital and scientific credibility they had previously developed into a powerful institutional space in the federal government. Through a chain of social contacts, Bush's recommendation for a wartime science policy agency was implemented by an executive order in June 1940, which created the National Defense Research Committee (NDRC), the predecessor to the Office of Scientific Research and Development (OSRD).

Over the course of the war, Bush and his colleagues developed close contacts with members of Congress and the business community. The technical achievements[22] of NDRC and later OSRD bolstered the credibility of scientists in general and Bush and his colleagues in particular. Bush became Franklin D. Roosevelt's informal science advisor and was ultimately able to persuade the president to request from him a report on postwar research policy. The report, *Science—The Endless Frontier* (1945), played an important role in defining the terms of postwar debate, and the social connections and credibility of Bush and his colleagues placed them perfectly to influence the legislative debate.

Here, the organizational configuration of society and, in particular, the organizational configuration of policymaking, broadly construed, affected the degree of power any one set of actors had in shaping policy outcomes (Hall 1986: 19). The establishment of the OSRD created the organizational framework within which important opportunities and constraints were produced and these affected the outcome of the struggle over the nature of the postwar research policy establishment.

If the war brought elite scientists under Bush to prominence and raised the issue of research policy in the postwar period, it is true as well that Senator Harley Kilgore's early interest in a postwar research policymaking agency came out of the wartime situation. With other populists and New Deal activists he was concerned about what appeared to be serious government and corporate failures in the effort to prepare for the war. Congressional investigation pointed to overlapping government bureaucracies and inadequate government planning. Congressionally sponsored studies indicated preventable resource shortages and restrictions of trade and the flow of information through improper use of patents. It was within this context that Kilgore made his first proposal in 1943 for a postwar research policymaking agency.

The war led to two opposing visions of the organization of federal research policy for the postwar period (see table 5.1). These contrasting proposals started the legislative ball rolling. Funneled through the state, it was nearly eight years and several distinct proposals before legislation was ultimately passed. During this period, the role for the proposed National Science Foundation became progressively narrower, and new and existing agencies claimed bits and pieces from the original proposals, leaving the Science Foundation with limited responsibilities. To understand the long, tortured, and twisted trajectory from Kilgore's initial proposal in 1942 forward, we must look at the structure of the state and civil society and the relationship between the two.

Science—The Endless Frontier marked a high point of elite scientists in a *formal role* shaping the trajectory of postwar science policymaking.[23] The war pushed the line dividing state and society further into civil society than it had been previously. Scientists and representatives of business were brought directly into the state, and, with a dense web of social connections between elite scientists and business and state elites, Bush and his colleagues were perfectly positioned to gain a prominent place in the state. These formal positions served Bush and his colleagues well, enabling them to strengthen *informal contacts* with members of Congress and the business community. In turn, these contacts enabled Bush and others to directly influence the course of legislation that ultimately led to the National Science Foundation.

On the one hand, with *Science—The Endless Frontier,* scientists played a direct formal role in shaping debate over the configuration of a postwar research policy agency. On the other hand, Bush's formal role as head of the Office of Scientific Research and Development enabled him to make a wide range of contacts both inside and outside the state, and he was able to use these contacts—in the context of a highly permeable state—to stop compromise legislation, ultimately giving other interests an opportunity to realize their own state-building projects and narrowing the role of the National Science Foundation when it was finally established.

Like scientists, representatives of business played both formal and informal, direct and indirect, roles in the debate over a postwar research policy agency. Business representatives served as what broadly might be termed state managers. For example, Oliver Buckley of Bell Laboratories sat on one of the committees that drafted Bush's *Science—The Endless Frontier* report in 1945. Buckley was also a member of the Directors of Industrial Research (DIR) group, an organization of the research directors of many of the country's leading science-based companies, and he sought the advice of other research directors in the organization concerning the questions that Bush's committees would be addressing. Here again is a clear case of the blurred line between state and civil society. While we do not know precisely what Buckley's influence was on the Bush committee, nor how much he was swayed by his research director colleagues, we do know that the central positions taken in the report are fully compatible with the views of business representatives who played an active role in the research policy debate.

The DIR also played an informal role in shaping one of the early bills to establish a national science foundation. In one case, Republican Senator H. A. Smith sought the advice of the research directors, and following their suggestions, Smith excluded support for applied research from the responsibilities of his proposed agency. This bill failed, the result of intragovernment squabbling and the lack of programmatic parties. But the bill that ultimately passed in 1950 did establish an agency in keeping with the demands and interests of elite scientists and science-based industry.

To understand the eight-year delay in passage of legislation to establish a postwar research policy agency in the United States—what came to be called the National Science Foundation—we must look at the organizational structure of the U.S. state and civil society. The permeable and fragmented character of the American state, combined with undisciplined political parties, led to one delay after another; two bills died in committee at the hands of cross-party coalitions, and one bill was vetoed by a Democratic president presented with a Republican piece of legislation.

A temporary change in the structure of the state, which limited the effects of permeability, and atomistic congressional politics, permitted passage of the legislation in 1950. When the final bill was signed eight years after Kilgore first proposed a single comprehensive science agency, many of the responsibilities Kilgore had envisioned for his agency had been taken up by agencies created in the meantime with narrower, highly circumscribed responsibilities. The Office of Naval Research became the lead agency supporting academic research in the immediate postwar period. The Atomic Energy Commission too ended up supporting basic university research, and the Joint Research and Development Board guaranteed the military a role in postwar research policy. In addition, the prolonged struggle to establish the National Science Foundation gave Public Health Service officials an opportunity to guarantee their organization's continued dominance of medical research.

Certainly, fragmentation and permeability beget further fragmentation. The delay in passage of national research policy legislation led to the construction of a fragmented matrix of science policy institutions, each with limited roles. In addition, the shape of National Science Foundation legislation—and ultimately the foundation itself—was significantly influenced by elite scientists and business representatives. Their roles were formal—directly inside the state—and informal, playing on a wide range of social connections and the permeable character of the state.

Plan of the Book

This book is divided into seven chapters. In chapter two, I provide a detailed description and analysis of the pre–World War II scientific field. In chapter three, I provide an in-depth look at the Office of Scientific Research and Development (OSRD) and science and scientists during World War II. In chapter four, I compare the legislative agendas of Harley Kilgore and Vannevar Bush. Chapter five draws on and extends work in political sociology and provides a detailed analysis of the legislative history of NSF legislation.

In chapter six, I discuss the implications of the five-year delay in passage of National Science Foundation legislation for the institutional geography of government research policymaking. In addition, to lend credibility to the account I provide of the genesis of the federal system for research policymaking after World War II, I undertake a cross-national comparative reconnaissance of the organization of states and societies and their relationship to the character of national research policymaking.

In chapter seven, I consider post-1950 efforts to establish research policy agencies at the federal level in the United States with broader policymaking and coordination mandates than the NSF has. I explore the failures of each and analyze the prospects for the establishment of a more comprehensive research policy (agency) during the current so-called crisis of competitiveness.

2

MAPPING SCIENCE

The Scientific Field in the United States, 1850–1940

> It is not just the people who work in the laboratories who do science, but everyone who takes part in sponsoring, producing, justifying, or making use of scientific knowledge.
>
> Robert E. Kohler, 1990

That science is a complex social system is often asserted by social scientists and opinion leaders. But however often it is asserted, it is nevertheless frequently overlooked. Understanding the complex and multilayered character of science and its historically specific form in the period prior to World War II is crucial to comprehending how struggles to redefine research policymaking and patterns of research funding played themselves out in the period after the war. Toward such an understanding, this chapter provides an analysis of what I will term the *scientific field*—an institutional map of science—in the period from about 1850 to about 1940. I consider the type of research undertaken, where it is being done, why it is being done, who is funding it, and why it is being funded.

The character of the prewar scientific field affected the (re)organization of research policymaking after World War II in at least four distinct ways: by shaping or defining a set of collective projects and interests for certain groups of scientists and business elites; by providing an institutional legacy and a policy history on which actors drew in specifying their vision of the postwar scientific world; by providing the basis for the development of a range of social connections—social capital—on which actors drew in their efforts to shape the postwar scientific field; and by providing a context within which scientists could develop the cultural capital or credibility—scientific authority—that facilitated their entree into government during the war.

Broad areas in the scholarly literature on science have explicitly or implicitly treated the realm of "knowledge production" as an autonomous and distinct social sphere.[1] As late as the early nineteenth century, research was the domain of self-supporting amateurs. Under these conditions, it may have made sense to speak of autonomy, at least in terms of resource dependence. Scientists had little contact with industry or government; research was not carried out primarily in universities, and foundations were absent from the scene. By mid-century, however, "the technical means of producing scientific knowledge increasingly outgrew the capacity of individual provision and control so that they had to be collectively organized and controlled" (Whitley 1984: 65; see also Restivo 1988: 211).[2]

Scientists became more dependent on external sources of support. Laboratory methods gradually displaced other approaches, and the more general move toward professionalization of science eventually resulted in "nearly all work being done by employees" (Whitley 1984: 48). Scientists moved from their independent laboratories behind the house to universities and sometimes to firms and government laboratories (Bruce 1987).

The dependence of university scientists on external sources for resources has increased dramatically since the mid-nineteenth century. Additionally, industry and the state have developed independent requirements for research. As a consequence, "intellectual priorities [have become] less totally controlled by purely academic interests and more open to resource allocation decisions in [nonacademic] . . . employment structures"; scientists' "claims to, and control over, the increasingly expensive laboratory facilities required for the highest reputations [have progressively become] . . . mediated by non-academic interests and goals" (Whitley 1984: 283–84).

The professionalized scientific field has been characterized by a progressive loss of this autonomy although, of course, the nature of autonomy varies by time and place. From the late nineteenth century through the present day, the American scientific field has been a shifting configuration constituted by the overlap and interaction of several institutional spheres: foundations, universities, scienced-based industry, and the state (Kleinman 1991). At some level, these spheres are distinct and possess independent logics and are composed of actors with distinctive projects and interests. But interaction often blurs boundaries in a range of ways. Interactions between industry and the university—close cooperation or collaboration—may define a project for university scientists shaped not simply by the logic of the university, but also by the logic of capital. Similarly, the world view adhered to by scientists is rarely unaffected by the values of "nonscientific" social spheres.[3]

Science and Academia

Research—whatever we mean by the term—has been, and continues to be, performed in many locations. But from its emergence in the late nineteenth century, the university has maintained its status as the primary locus of scientific research (Kuznick 1987: 11). The American research university has been, and remains, a major source for making one's livelihood from scientific investigation. The university has progressively come to dominate the production and validation of knowledge—and even where there are opportunities for scientific employment outside the university, university employment retains uniquely high status in the area of knowledge production (Whitley 1984: 66). Thus, it seems appropriate to begin the empirical portion of this chapter with a discussion of research universities in the United States from their emergence in the mid- or late nineteenth century until the onset of World War II.

The prewar university affected the reorganization of postwar research policymaking and funding patterns in at least three ways. First, the links established between science-based industry and university scientists, especially during World War I, provided space for the genesis of a political agenda that linked the interests of business and those of elite scientists. Second, a number of important institutional and policy legacies were established. The university became recognized as the locus of basic science. The legitimacy of concentrating research resources within a few universities was accepted, the principle of elite scientist control of science was established, and calls for a central federal science agency became commonplace as university scientists, industry, and the government cooperated during World War I. Finally, the university was a place where scientists could develop their cultural capital, and the value of this capital became clear to those outside the university when university scientists helped develop technologies used in the war.

Typically historians date the genesis of the American research university at sometime after the Civil War. Prior to the late nineteenth century, higher education in the United States focused on general rather than specialized training and exposure to "liberal culture" and the "classics" (Gruber 1975: 10–12; Bruce 1987: 326, 327). Professors were primarily teachers, not researchers. With the introduction of elective courses and the growing acceptance of specialized training, "science expanded its academic presence" (Bruce 1987: 327).

It was in the 1870s that research-based graduate studies became a permanent component of American higher education (Geiger 1986: 9). In the late 1860s, Josiah Willard Gibbs, the first internationally important American theoretical scientist since Benjamin Franklin, was awarded a doctorate by Yale, an institution that led the way in establishing the research model of graduate study (B.L.R. Smith 1990: 21). But it was not until a decade and a half later that Johns Hopkins University was founded, and it was Hopkins "far more than any other contemporary American university [that] actively encouraged original investigations by faculty" (Geiger 1986: 8).

If the founding and recognition of the centrality of American research universities date from the mid- or late nineteenth century, the origins are earlier. The American research university was modeled in important ways on the great universities of Europe. Beginning in the early nineteenth century, American students went to German universities "to acquire the professional and advanced education that was not available at home" (Gruber 1975: 17). Scholars from such places as Harvard and New York University returned from European research trips with visions of organizational reforms and of the possibility of creating universities with adequate research facilities and sufficient freedom from teaching to undertake research (Curti and Nash 1965: 108).

Surely, then, part of the emergence of research universities in the United States must be attributed to a mimetic process—a process of institutional imitation.[4] Of course, realizing visions of a true research university required resources. In the nineteenth century, tuition proceeds—money from instruction—helped support research (Geiger 1986: 2, 3). In addition, ad hoc philanthropy was an important source of research support for private universities prior to World War I, and state universities relied on state government funds (79). Noninstitutional support—support from individuals—came in great and small amounts. Universities initiated fund-raising efforts after the turn of the century, and alumni were among early supporters of university research, as were local communities (84, 43). Foundations, too, were an important source of research support prior to World War I (234).

Of course, large individual donations in support of research or university development more generally were typically provided by a business elite, which made its fortune as the United States moved rapidly to industrialize (Curti and Nash 1965). The great foundations, too, were created by leading industrialists such as John D. Rockefeller and Andrew Carnegie. And Rockefeller himself provided the funds to help get the University of Chicago off the ground (Geiger 1986: 197; B.L.R. Smith 1990: 21). If few "strings" were attached to this support, it is important to note that even at this early date universities, and thus scientific research, were intimately linked to the economy, and the interests of universities were intimately tied to the fortunes of prominent capitalists.[5]

It is hard to say precisely what type of research was undertaken in American research universities from the late nineteenth century until immediately before World War II.[6] At the ideological level at least—in terms of what scientists saw themselves doing—during the period under discussion, the university was dominated by a commitment to basic or fundamental research. At the same time, such research was not the only work undertaken in American universities during the period from the late nineteenth century until World War II. The most obvious early exception to this basic research orientation is the investigation undertaken at land grant universities.

Established in 1862 by the Morrill Act, land grant universities were intended to support research of use to farmers and the rural population.[7] Passage of the act occurred within a context of growing dissatisfaction with higher education and the emergence of agricultural organizations devoted to advancing the interests of farmers (Axt 1952: 37).[8] The act gave each state 30,000 acres of public land or the equivalent in scrip, and states were expected to sell the land or scrip and use it to fund an agriculturally oriented research university (41–42). In 1887, the land grant colleges were strengthened by passage of the Hatch Act, which provided federal support for the establishment of agricultural experiment stations and thus a context for the systematic application of scientific research to agricultural problems (U.S. House Task Force on Science Policy 1986a: 9–10; Kloppenburg 1988).

Prior to World War II, outside of agricultural research undertaken at land grant universities, the federal government sponsored little university research. Most nonagricultural research sponsored by the government was undertaken by federal agencies and in federal laboratories. The federal government's role, however, was radically transformed after World War II. In 1940, the government sponsored only about $13 million of research in universities throughout the country, and most of this was provided by the U.S. Department of Agriculture (USDA). By contrast, in 1950 the government spent about $150 million for research in U.S. universities, and this was distributed by more than a dozen federal agencies (Axt 1952: 86–87; Geiger 1986: 60; Kevles 1988a: 119).

As I explained in chapter one, this transformation can be attributed to the interaction of several factors—World War II and the ensuing Cold War, the structure of the American state, and, of course, less directly the structure of the prewar scientific field and the mobility project of a group of elite science administrators. I will return to a detailed discussion of these in later chapters.

If the federal government in a limited role—through the USDA—was the prime sponsor of so-called nonbasic directed research in American universities, it was not the only sponsor. Industry support for university research in the interest of profit dates back to early in the nineteenth century when firms occasionally employed university researchers. But such cases were rare during the first part of the nineteenth century (Noble 1977: 111). Much later, in 1902, the Massachusetts Institute of Technology (MIT) developed close links with American Telephone and Telegraph (ATT), which provided regular research support for the Institute's Department of Electrical Engineering (Geiger 1986: 177). In 1908, MIT created the Research Laboratory of Applied Chemistry and became "the first academic unit dedicated to performing research for industry" (Geiger 1989: 3).

But it was during World War I that the relationship between America's research universities and science-based industry was solidified. At the onset of the war, a group of academic and industrial science elites created the National Research Council (NRC) as a spinoff of the National Academy of Sciences. The explicit aim of the council was to advise the government on matters of science during the war. Indeed, on the basis of support from the Rockefeller and Carnegie Foundations, along with some federal funds, the NRC established a committee that informed university scientists about projects and problems in which the government was interested (Axt 1952: 78). On behalf of the state, the NRC forged a link between university and industry and helped achieve large-scale production of optical glass, nitrates, and poison gas (Geiger 1986: 97).

If the NRC was an important force in the war effort, its strategic military importance is perhaps overshadowed by the institutional precedents it established. For industry, it marked the beginning of truly close ties between universities and industry. In the 1920s, firms eagerly hired university-trained scientists and technicians. Some companies also provided university fellowships to expand the scientifically literate workforce. Still other firms hired university scientists as consultants (Geiger 1986: 175). Businesses also provided direct research support to university scientists willing to work in areas of interest to funding corporations.

But World War I and the policies of the NRC also established a number of institutional legacies of relevance to the postwar period. First, they created tight links between universities, private industry, and philanthropic foundations. This overlap and interaction became the taken-for-granted means of organizing science in the period after the war; in addition, these kinds of institutional connections made possible the social contacts between industry representatives and science administrators that would help to define the agenda of Vannevar Bush and his colleagues in the postwar period. As well, Roger Geiger suggests, the NRC set an important pattern for the organization and direction of post–World War I U.S. science: "That *a central agency* was desirable became an axiom of the ideology of [American] science" (1986: 99 emphasis added). Again, calls for a central research policy agency came from the lips not only of Bush and his colleagues, but also from his New Deal opposition, particularly Harley Kilgore. Furthermore, the establishment of the NRC during World War I "consecrated" the direction of science policy in the hands of a small elite (101). By World War II, this precedent of elite science control had become firmly entrenched, and it was a small elite that defined scientists' agenda and ultimately the nation's agenda for research policy after 1944.

The legacies established during World War I—which were to play such an important role in the reconstitution of the scientific field, particularly the organization of research funding—did not fundamentally alter the nature of support for university research during the interwar period. A significant portion of university budgets in the period between 1919 and 1940 came from student fees in private schools and from state appropriations for public universities (Weart 1979: 312), and foundations continued to be important supporters of university research (Geiger 1986: vii).

Support for university research during the interwar period also created a pattern of concentration that was reinforced and legitimized during World War II and became an issue of some controversy in the post–World War II period, as conflicting groups struggled over the reconstitution of research policymaking and the scientific field. Between 1902 and 1925, the Rockefeller General Education Board "awarded almost two thirds of its total grants for science to just eight institutions—Caltech, Princeton, Cornell, Vanderbilt, Harvard, Stanford, Rochester, and Chicago" (Kevles 1987: 192). In the late 1930s, it was a similarly small group that dominated university research. Among the universities receiving two-thirds of all Rockefeller General Education Board support in the first quarter of the twentieth century, only Vanderbilt and Rochester were not among the major spenders on research in the 1930s (see tables 2.1 and 2.2). And of the sixteen universities spending the most on research in the 1930s, all but four were among the top twenty-five nonindustrial contractors to the Office of Scientific Research and Development, Vannevar Bush's wartime government science agency (see tables 2.1 and 3.3). Concentration of support developed early in the century had become firmly established by the 1930s and was reinforced by government practices during World War II.

Foundation(al) Support for Science

As I have noted, the early development of university-based scientific research depended heavily on support from foundations, and two founda-

Table 2.1 Estimates of Total University Research Expenditures in the 1930s

More than $2,000,000
University of California, University of Chicago, Columbia University, Harvard University, University of Illinois, University of Michigan
$1,500,000–$2,000,000
Cornell University, University of Minnesota, University of Wisconsin, Yale University
$1,000,000–$1,500,000
Massachusetts Institute of Technology, University of Pennsylvania
Under $1,000,000
Johns Hopkins University, Princeton University, Stanford University, California Institute of Technology

Source: Geiger 1986: 232.

tions in particular: the Rockefeller Foundation and the Carnegie Endowment. These foundations were established at roughly the same time and in the same context that saw the emergence of universities as major spheres of scientific research. In this section, I consider the role of foundations in the prewar scientific field. I explore the genesis of American foundations, who controlled them and with what effects, and the legacy foundation support created for the postwar scientific field in the United States.

Foundations and foundation funding of university research shaped postwar research policymaking and thus the postwar scientific field in several ways. First, by concentrating research resources, foundation support helped develop the research capacity of a few universities and established an important policy legacy: the principle of concentrating resources to promote the "best science." This approach became a major point of controversy in the debate over postwar government research funding. Second, foundations created an important precedent by supporting university research by contract. The government followed this

Table 2.2 Early Rockefeller Foundation Support for Natural Sciences Research in U.S. Universities

Recipient	Amount ($1,000s)
Support Recipients, 1929–33	
Cal Tech	600
Harvard	595
Johns Hopkins	428
Chicago	165
Princeton	116
Ohio Wesleyan	20
Minnesota	15
MIT	14
Alaska Agricultural	10
General Education Board Appropriations, 1923–31	
Cal Tech	3,079
Princeton	2,000
Chicago	1,798
Rochester	1,750
Cornell	1,500*
Harvard	1,175
Stanford	870*
Vanderbilt	693
Yale	500
Texas	65
North Carolina	15
Columbia	10

Source: Kohler 1990: 202, 256.
Note: I have excluded nonuniversity award recipients from this table. An asterisk includes pledges not redeemed.

approach to supporting university research during World War II, and this mechanism, which gave university scientists considerable autonomy, was established in spirit in postwar government funding of research. Third, foundations played an important part in developing the social network established between scientists, government officials, business leaders, and foundation managers during World War I. This network was crucially important in defining the post–World War II policy agenda of elite scientists. Fourth, in several ways foundation support for scientists bolstered the broad social authority of scientists. Finally, by creating the basis for big science, foundations contributed to elite scientists' dream of vast resources, a dream that guided their collective advancement project.

The Rockefeller Foundation and Carnegie Corporation were established two years apart, the Carnegie in 1911 and the Rockefeller in 1913. In the years immediately after World War I, these two foundations dominated philanthropy for scientific research, and Roger Geiger argues that their establishment "permanently changed the landscape of American philanthropy" (1986: 143). In 1940, fully one-third of total research expenditures by foundations was provided by the Carnegie Corporation and the Rockefeller Foundation (U.S. Senate 1945a: 38).

We can analytically distinguish the ideological and material origins of American foundations. Nielsen speaks to the ideological origin:

> The establishment of the very large American foundations, which began in the late nineteenth century, represented the beginning of a whole new phase in the American tradition of charity and altruism. Up until roughly the time of the Civil War "benevolence" and "good works" were interpreted by the wealthy both in the East and the Middle West to involve obligation of personal service and stewardship. (Nielsen 1989: 8)

New ideas about charity were ultimately fused with the Progressive ideal that "human welfare was best promoted by the systematic and rational application of 'objective' knowledge" (Kohler 1979: 251). This occurred within a context in which the role for government was seen as limited and marked off not only from religion but from "broad areas of welfare, medicine, science, culture, and much of education," which were believed to belong to the domain of the private sector. Ultimately, the idea of charity through personal service, rooted in the Judeo-Christian tradition, was monetized and bureaucratized into a new institutional form: the foundation (Nielsen 1972: 379; 1989: 8).

As to the material origins, such new ideas of philanthropy could only be put into practice where there was money, and it was in the period of

American industrialization and economic prosperity of the late nineteenth century that the wealth that made the great foundations was acquired. The early foundations "were based on Eastern [U.S.] fortunes gained from basic industries and natural resources: oil, steel, and coal"—the money of a rising economic elite, not an established aristocracy (Nielsen 1989: 10, 13). Andrew Carnegie became a powerful steel magnate, and "in the closing quarter of the nineteenth century, by a series of complex combinations and mergers, he made his coal-and-steel complex so powerful that it had no effective competition." John D. Rockefeller created the Rockefeller Foundation with $50 million in shares from his Standard Oil Company of New Jersey (Nielsen 1972: 32, 50).

Early on, the philanthropic uses of the fortunes of the likes of Andrew Carnegie and John D. Rockefeller were, indeed, under the control of these affluent industrialists. During the late nineteenth century Rockefeller donated money in a haphazard fashion to a range of causes, mostly related to his interests in the church. But in 1892, Rockefeller hired Frederick Gates, a former Baptist minister, as an advisor on philanthropic matters, and this marked the gradual transformation of Rockefeller philanthropy from direct Rockefeller control to control by a professional managerial group (Nielsen 1989: 84–85).

Although the timing was different, the story was quite similar in the case of the Carnegie fortune. According to Curti and Nash, "until Frederick Paul Keppel was elected president of the [Carnegie] corporation in 1922, it was little more than a device for Carnegie to gain assistance in his personal giving to libraries, churches, and other philanthropic enterprises" (1965: 223).

While foundation boards typically remained controlled by capitalists or their aparatchiki (Nielsen 1972: 385; Nielsen 1989: 19), direct control of foundation funds in the period before the war came to be exercised, to a significant degree, by a cadre of foundation managers whose actions may have been constrained by capitalist ideology but who were nevertheless engaged in independent projects to realize their own vision of the scientific enterprise (Kohler 1990). Indeed, the independence of these managers was often guaranteed by the foundation founder in the document establishing the foundation (Curti and Nash 1965: 213). In the case of the Rockefeller Foundation, Nielsen contends that the technical or scientific expertise of the foundation's division heads and their reliance on a separate board of scientific advisors made their recommendations to the board irresistible (1972: 58). And according to Richard Whitley, by the 1920s the independence and control exercised by foundation managers at the major American foundations "enabled a relatively small group of officials and their advisors to influence directly the production of knowledge in favored areas" (1984: 284). Whitley suggests further that "by funding research in particular areas directly, intellectual priorities could be altered and co-ordinated by a central group which acted across universities" (284).

Prior to 1920, foundation support centered on research with practical applications consistent with foundation commitment to improving human welfare (Coben 1979: 232; Geiger 1986: 140). Support was most often made to establish independent research institutes, typically under foundation control (Geiger 1986: 142; Kohler 1987: 135). In addition, funds were contributed to individual university endowments to be used in any way the recipient institution desired (Coben 1979: 232).

The waning days of World War I marked a shift in foundation policy. Among other things, in 1919, the National Research Council—the wartime organization linking universities, foundations, the business community, and the state—created a postdoctoral fellowship program with money provided by the Rockefeller Foundation. These fellowships, according to Whitley, enabled U.S. theoretical physics to blossom and form a national elite for physics (1984: 285). Rockefeller also provided fellowships of its own (Coben 1979: 232). Such support strengthened the research focus of developing American universities by giving scientists time away from teaching.

In addition to providing individual fellowships, after 1920, foundations increasingly targeted their support within universities. Foundation funds went to specific projects rather than to general support for university endowments. According to Curti and Nash, this shift also enhanced the research function of American universities. Almost all the projects supported "tended to direct higher education away from its traditional teaching function and to give emphasis to scholarly investigation and publication" (1965: 223).

The Rockefeller Foundation's move to support specific projects was stalled by a fear of having the organization criticized for directing the course of scientific research. But in 1928, the foundation was reorganized into broad research divisions, and the focus of grants turned to individual research projects (Kohler 1979: 255). Support was generally guided by the belief that scientific research was the "surest means of advancing civilization and the well-being of mankind" (Nielsen 1972: 55–56)—that scientific research, in the words of Wickliffe Rose, one-time head of Rockefeller's General Education Board, is "*the* method of knowledge" (quoted in Nielsen 1972: 55–56; emphasis added). And support for research was concentrated "within the few strongest university departments of science" (Coben 1979: 234–35). The aim of Rockefeller Foundation support was to "make the peaks higher" by supporting research at institutions already doing the best work (Geiger 1986: 161).

Importantly, both the claim that scientific research is *the* method of knowledge production and the decision to support primarily the leading research institutions shaped the organization of research in the postwar period. Rockefeller support helped create the preeminent research institutions in the United States and legitimate the idea that they, and not "second-rate" schools, should be supported. In addition, the rhetorical power granted scientific research was drawn on to strengthen the cause of the scientist-institution builders in the period during World War II and beyond.

Beyond the legacy established by Rockefeller for the period after the war, the Rockefeller Foundation provides a good case study for exploring Whitley's claim that foundation managers in the post-1920 period directly influenced the direction of knowledge production. The early shaping of biology by Rockefeller's Warren Weaver is an interesting case in point. Weaver, a classical physicist and former mathematics department head at the University of Wisconsin, became director of the foundation's natural sciences division in 1932. Provided with a great deal of institutional autonomy, Weaver shifted the focus of foundation support from providing general research capital to supporting specific research. This research-specific support took the form of three-year grants for planned or programmatic projects. This shift in funding mechanism enhanced Weaver's control of priority setting; moreover, drawing on his physics background, Weaver shaped an agenda that relied heavily on providing support to projects in the biological sciences that applied the techniques and discoveries of the physical sciences (Geiger 1986: 165; Kohler 1979: 273–74; Kohler 1990: 265–394).

In addition to the role of Weaver and the Rockefeller Foundation in shaping the contours of interwar biology, the Rockefeller Foundation played a central role in shaping big physics in the United States. In the spring of 1940, the foundation granted Ernest Lawrence over a million dollars for a cyclotron. The new physics signaled by this grant relied both on theorists and experimentalists. It required massive machines and large organizations (Kevles 1987: 285–86). The mark of foundation support on the new physics is indicated by the fact that by 1940 about twenty-five percent of all papers that appeared in the *Physical Review* were acknowledging outside support for fellowships and grants, chiefly from foundations (Weart 1979: 313). Significantly, Lawrence's cyclotron played an important role in the development of the atomic bomb in the early years of World War II, and, in fact, twenty-three members of the Manhattan Project were, in early years, Rockefeller Fellows (Nielsen 1972: 62).

In general, Rockefeller was a dominant force in shaping science in the interwar period. In 1934, the foundation provided thirty-five percent of all foundation support for research and seventy-two percent of all support in the natural sciences (Geiger 1986: 166). Beyond shaping the contours of research fields, Rockefeller support is credited with facilitating the development of new research tools and techniques, including spectroscopy, x-ray diffraction, chromatography, and the use of tracer elements (Nielsen 1972: 62; Kohler 1990: 358–94).

The development of foundations and foundation support of scientific research in the period before World War II played an important role in shaping the contours of research policymaking and ultimately the scientific field in the period after World War II. Foundations had a central role in developing the research capacities of American universities, and universities became the crux of the research system in the United States. The Office of Scientific Research and Development (OSRD)—the wartime government science agency headed by Vannevar Bush—followed the standard set by foundations in the period after World War I of providing research for specific projects and permitting scientists to undertake this research at their home universities. This pattern was accepted early on in debates about how the federal government should support research in the post–World War II period. Concentration of research support was a direct result of foundation decision making during the interwar period, and this precedent, too, was accepted by OSRD, although it was a point of contention in debates over the post–World War II organization of science.

As I noted in the previous section, the creation of the National Research Council during World War I facilitated the creation of a social network between business, science, and foundation elites. This network proved important in defining research policy during World War II and ultimately the postwar period. Moreover, foundations were instrumental in the NRC's creation. The Rockefeller Foundation supported the council's fellowship program, and the Carnegie Corporation supported the NRC with a $5 million grant for a permanent home and an endowment (Geiger 1986: 147).

In a circuitous fashion, foundations also enhanced the cultural capital of science generally. They reinforced a discourse of science as a superior way to know, and they supported the development of a peer review system premised on the claim that scientists were the most qualified to determine the most promising lines of research. This was an ideology that supporters of a non-expert-controlled research policymaking system could never really challenge. Foundations further indirectly enhanced the cultural status of scientists by supporting the development of the nuclear physics that helped "win" the war. Finally, in establishing the basis for big science—physics in particular—foundations may have indirectly shaped the demands and interests of scientists in the nature of funding postwar research. An office and a pencil would not be enough. Big physics demanded big bucks and ultimately big government support.

Science and Industry

Industry concerns with research in the period before World War II also played an important role in shaping the contours of postwar research policy. By promoting links between academic scientists and industry, prewar industrial research helped define the policy agenda that guided the collective advancement project of Bush and his colleagues. In addi-

Table 2.3 U.S. Industrial Research Expenditures, 1920–40

Year	Expenditure (in $1,000s)
1920	$ 29,468
1925	64,000
1930	116,000
1935	136,000
1940	234,000

Source: Bush 1945: 86.

tion, the expansion of industrial research during the period shaped demands by research-based industry for the exclusion of applied research from financial support of the government (table 2.3).

Through individual donations to universities and the establishment of philanthropic foundations, industrialists played an important *indirect* role in shaping the configuration of pre–World War II research and research policy in the United States. In addition, leaders of industry had a direct interest in science. Prior to the turn of the twentieth century, "industrial research remained essentially the unorganized effort of individuals" (Noble 1977: 111; see also U.S. House Task Force on Science Policy 1986a: 10). Independent chemical laboratories served individuals and firms. Consultants from independent laboratories and academia rendered services to industrial concerns; indeed, a few firms had laboratories that undertook testing and were responsible for product innovations, but, in general, few firms had independent research capacity, and those that had such capacity did not have systematic research and development programs (Birr 1979: 195–96).

The systematic use of research by industry for product and process improvement and development, not merely testing, emerged very late in the nineteenth and in the early twentieth centuries (Noble 1977: 5). The internalization of research aimed at the "systematic application of scientific knowledge to the process of commodity production" occurred within the context of bitter market competition and is seen by many historians as a means by which firms sought financial security and stability (Noble 1977: 5; Birr 1979: 179; National Research Council 1940: 19). But, in addition, in-house research capacity required a certain threshold of financial resources and, not surprisingly, like the genesis of America's foundations, the emergence of industrial research coincides with the "period of consolidation that followed the birth or youth of America's corporate giants" (Wise 1980: 412; see also Noble 1977: 111). If consolidation provided the economic resources for large firms to develop serious research capacity, size and complexity also may have made the reliance of industry on random discoveries less useful (Perazich and Field 1940: 41).

Indeed, "most of the early laboratories appeared in industries such as electricity or chemicals where there were large firms with sufficient financial resources and stability to support the laboratories and where there was a rapidly changing, competitive technology which made successful research imperative for the sponsoring firm" (Birr 1966: 68). In 1920, approximately two-thirds of the research workers recorded in the first industrial research survey conducted by the National Research Council were employed in the electrical, chemical, and rubber industries (National Research Council 1940: 34). General Electric established its research laboratory in 1900. DuPont established its first research laboratory in 1902, and American Telephone and Telegraph set up a research lab in 1904 (Birr 1979: 199).

If it was large firms that initially were able to internalize research capacity, research was highly concentrated in large firms even as late as 1938. Although as many as fifty-two percent of all industrial concerns may have considered research a company activity in 1928, a study by the Temporary National Economic Committee found that "thirteen companies—less than one percent of those reporting research activities in 1938—employed more than one third of all research workers. One half of industrial-laboratory personnel were employed by forty-five large laboratories, all but nine of which were owned or controlled by companies among the nation's two hundred leading non-financial corporations" (Noble 1977: 111, 120) (table 2.4).

Where research was undertaken in small firms—firms with fewer than 1,500 employees and assets no greater than $2.5 million in 1940—the focus of research was " 'organized fact finding' of an immediate and practical" nature. Research was not necessarily set apart as a functional

Table 2.4 The Forty-Five Companies Employing Half of Total Research Personnel in 1938

Aluminum Company of America	Hudson Motor Car Company
American Can Co.	Humble Oil & Refining Co.
American Cyanamid Co.	International Harvester Co.
Atlantic Refining Co.	Linde Air Products Co.
Bakelite Corp.	Monsanto Chemical Co.
Bell Telephone Laboratories	Pennsylvania Railroad Co.
Chrysler Corp.	Philco Radio & Corp.
Consolidated Edison Co. of New York	Pittsburgh Plate Glass Co.
Crucible Steel Co. of America	RCA Manufacturing Co.
Dow Chemical Co.	Republic Steel Corp.
duPont, E.I. de Nemours & Co.	Shell Development Co.
Eastman Kodak Co.	Sinclair Refining Co.
Electric Auto Lite Co.	Socony-Vacuum Oil Co.
Firestone Tire & Rubber Co.	Standard Oil Co. (Indiana)
Ford Motor Co.	Standard Oil Co. of California
General Electric Co.	Standard Oil Co. of Louisiana
General Motors Research Corp.	Standard Oil Development Co.
General Motors Truck & Coach Co.; Yellow Truck & Coach Manufacturing Co.	Sun Oil Co.
	U.S. Shoe Machinery Corp.
	U.S. Rubber Products Co.
B. F. Goodrich Co.	Universal Oil Products Co.
Goodyear Tire & Rubber Co.	Western Union Telegraph Co.
Gulf Research & Development Co.	Westinghouse Electric & Manufacturing Co.
Hercules Powder Co.	

Source: Works Projects Administration 1940: 68.

unit; that is, these firms did not necessarily have distinct laboratories. By contrast, large firms typically engaged in research of a "continuing and more intensive nature, carried on in specially organized departments or laboratories and encompass[ed] in certain instances advanced research which the small company can rarely afford" (National Research Council 1940: 80).

In the period before World War II, the structure of industrial research was dualistic in a double sense: on the one hand, there was the difference between the research undertaken in small and large firms; on the other hand, there was research undertaken within firms and research undertaken for firms in independent laboratories. In the period between 1900 and 1940, nearly 350 independent laboratories were established in the United States. They undertook contract research of a generic character, exploiting little or no firm-specific knowledge. In addition, they undertook research aimed at improving specific production processes employed by many firms, as well as the analysis of input quality (Mowery 1983: 355). Surprisingly, these independent laboratories were not used primarily by firms without autonomous research capacity. Instead, in the period between 1910 and 1940, "it is clear that for a substantial proportion of research clients, the independent research organization are complements to, rather than substitutes for, in-house research" (361). In the period between 1910 and 1919, just over twenty percent of the firms contracting with the Mellon Institute had in-house research capacity, but the percentage jumped to over fifty percent in the period between 1930 and 1940 (361). According to Mowery, "contract research was of limited efficacy as a substitute for an in-house research facility" (363). This is so, Mowery contends, because the most complex work contracted out required that firms have in-house capacity to utilize the contracted research, and firms without such capacity when they contracted for research typically required technically simpler work from the contractor than did the firms with in-house capacity (363).

Chemists and engineers dominated industrial research in the period before World War II. As late as 1938, these two groups constituted fifty percent of total personnel in industrial laboratories (Perazich and Field 1940: 11; National Research Council 1940: 13). The motivation behind industrial research—increasing profit, whether through new product or process development, quality improvement, or cost reduction—meant (and, indeed, continues to mean) that the organization of research in companies is different from that in a university setting. Noble points out that, "whereas the university researcher was relatively free to chart his own paths and define his own problems . . . , the industrial researcher was more commonly a soldier under management command, participating with others in a collective attack on scientific truth" (1977: 118). In an industrial environment, "scientists will tend to study topics which are irrelevant to reputational considerations and/or which require skills, knowledge, and procedures from a number of intellectual fields and so transcend reputational boundaries" (Whitley 1984: 51).[9]

Still, even if the motivations and organization behind industrial and university research were relatively distinct in the period before World War II, there was clearly overlap. Industrial scientists were, of course, trained largely at universities, and Whitley contends that the expansion of university physics teaching posts in the United States in the 1920s depended to an important degree on the growth of industrial jobs for physics Ph.D.s (1984: 284). In addition, even in the early history of industrial research, as I have noted, consulting by professors for industry was not uncommon. Finally, scientists who began their careers in the university might make the transition to industrial posts, for while industry could not offer the professional prestige of a professorship, it could sometimes offer superior research facilities (Wise 1980: 413).

As it was for universities, World War I was decisive in shaping the profile of industrial research. First, as I noted above, the war led to the collaboration of industrial research concerns with universities and the government, most notably through the National Research Council. But, more generally, according to Birr, "if the industrial research laboratory had its beginnings in the first decade of the twentieth century, it was not firmly established in this country until after World War I" (1966: 69). Before the war the United States was reliant on German dyes, scientific instruments, and optical glass. With U.S. access to these materials blocked by British blockade, industrial and retail consumers turned to domestic industry. Within this context, then, the modern American chemical industry became firmly established (Birr 1966: 62; Kevles 1987: 103; Noble 1977: 16).

With the centrality of industrial research for national economic independence and stability firmly established by World War I, industrial research grew dramatically during the interwar period. In 1920 there were about 300 industrial research laboratories in the United States. By 1930, there were some 1,400 labs, and the number had increased to 2,200 by 1940 (Kuznick 1987: 10; National Research Council 1940: 19).[10] The number of research personnel increased dramatically between 1920 and 1940, jumping from only 10,000 to nearly 60,000 (National Research Council 1940: 174).[11] Chemists especially benefited from the postwar growth of industrial research, and some seventy percent of chemists worked in industry from the end of the war through 1925 (Kuznick 1987: 10). As for spending, the best estimates suggest an increase from somewhere between $100 and $160 million in 1930 to $234 million by 1940 (Birr 1979: 200).

The genesis and presence of industrial research in the period prior to World War II perhaps played a less dramatic role in shaping the contours of postwar research and research policymaking in the United States than did foundations. Industry played little role in establishing universities as the pivot of the scientific field. The role of industrial research was, nevertheless, significant. During World War I, the importance of science for national economic autonomy and stability was established. In addition, the influence of industry in the overall scientific field made it certain that scientists rarely would be constituted simply as scientists. Instead, the overlap and interaction between academic and industrial science resulted in a discourse of science that was fundamentally fused with economic values. Part of the philosophy driving the project led by Vannevar Bush to control research policy stressed the centrality of basic scientific research for economic development; university research would thus become the basis for industrial science. In addition, Bush and his colleagues pushed patent policies vis-à-vis science that were strongly supported by industry, though opposed by New Dealers. Finally, of course, the overlap between university and industrial science and the government, especially through the National Research Council, helped create a network of social connections, which played a decisive role in shaping the scientific field in the postwar period. As well, the internalization of research by large-scale capital helped define the interests of business in a federal postwar research policy that restricted government funding to support for basic research.

Science and/for the State

Prior to World War II the federal government's role in the scientific field was relatively limited. Nevertheless, as a component of the prewar scientific field in the United States, the federal government played an important part in shaping postwar federal research policy and financial support for science. The prewar history of failed attempts to establish a *central science agency* was an important negative legacy; it foretold of things to come—a fragmented federal system for making research policy in the postwar period. In addition, the federal government, following the practices of the foundations, used an arm's-length system of contracting for university research and thereby helped legitimize the principle of granting government-funded researchers considerable autonomy. The right to such autonomy—an expression of scientific authority and a defining aspect of the collective advancement project of elite scientists—was further reinforced by World War I agencies that granted scientists decision-making authority.

In the first years of the new American nation, the federal government had little connection to, or involvement with, scientific research. A strict interpretation of the U.S. Constitution limited government involvement in science to the granting of patents (Dupree 1957: 14). The government did support some topographical research to benefit commerce and the military, but "as the eighteenth century and the first decade of the Constitution drew to a close, the new government had few tangible accomplishments in science and had made little headway in developing permanent institutions either to use science in its own operation or to disseminate it among the people" (19).

The early nineteenth century saw the expansion of the Coast Survey and government research to develop charts and maps for use by commercial interests. In addition, the needs of the U.S. Navy and demands of commercial interests "led to the promotion of science through the establishment in 1830 of the Depot of Charts and Instruments, and in 1842 of the Naval Observatory" (Lasby 1966: 254).

The exploration of the west was initiated in 1803 with the Lewis and Clark expedition. The expedition "blended the aims of empire and the desires for trade with the interests of science—the collection of flora and fauna, the acquisition of information concerning Indians, and astronomical observations" (Lasby 1966: 252). In the period before the Civil War, the Patent Office was also an instrument of what might broadly be termed science—or technology—in the interests of industry. The office maintained mechanical models and environmental materials illustrating the geology of the country (Dupree 1957: 47).

The Smithsonian Institution was launched in the mid-nineteenth century (Bruce 1987: 187–200; Dupree 1957: 66–90). Created with a bequest to the federal government by a wealthy Briton, a great deal of controversy surrounded its establishment. Some wanted to use the bequest to establish a national university; others advocated using the funds to build an observatory, and still others favored a national laboratory. As a consequence of this division, the institution's enabling legislation was vague, and it was left to the Smithsonian's first secretary, Joseph Henry, to define the institution's direction. Henry shaped an organization that focused on the promotion of research through publication and exchange, and, despite Henry's personal opposition, maintenance of research collections.

The Civil War led to increased government involvement in support for research. During the war, the navy supported military research—improving arms, ships, and steam engineering, as well as other military-related technologies (Dupree 1957: 120–25). A major development of the Civil War period was the creation in 1863 of the National Academy of Sciences. Chartered as an organization for consultation to the government on scientific and military matters during the war, the NAS was in addition "an attempt by prominent American scientists . . . *to centralize* control over American science" (Noble 1977: 150; emphasis added). Although a government creation, the charter creating the academy established the principle of scientist self-governance by permitting the organization's members to make their own rules and govern themselves. As well, some members of the academy attempted to establish the organization as *the* science advisor to the government (Dupree 1957: 139). The effort of scientists to create a self-governing *central science agency* was part of a legacy that infused the discourse and practice of science policymakers during and immediately after World War II.

The United States Department of Agriculture (USDA) was another prominent development of the Civil War period. According to Dupree, the department was an explicit challenge to concerns about constitutional limitations on government support for science. USDA established the principle that the government had the right to collect taxes and expend tax revenues for scientific research in the national interest (Dupree 1957: 151). The explicit aim of the department was to provide research support for farmers. With the establishment of land grant universities and state experiment stations, those supportive of creating a Department of Agriculture hoped to couple research and education (Dupree 1963: 457–58).

The nineteenth century also saw the first major effort in the United States to create a single, strictly government, science agency to *centralize* science policymaking. In 1884, the Congress created the Allision Commission to investigate existing science agencies in light of growing criticism on Capitol Hill of the ineffectiveness and inappropriateness of much government-supported research. The commission asked the National Academy of Sciences to undertake a study on the feasibility and appropriateness of creating a single department of science to undertake research not conducted in universities or by the private sector. While the 1886 report favored creation of such an agency, the commission decided against promoting a department of science on the grounds that it was not in the national interest and was not politically feasible (Dupree 1963: 461; Kevles 1987: 51; U.S. House Task Force on Science Policy 1986a: 7).

Federal scientists supported the creation of such an agency, believing that it would provide them with self-determination. Scientists outside the government, on the other hand, opposed calls for what they saw as political interference in science. The debate over the Allision Commission and its report raised a question that was to be central to the debate over the organization of postwar research: "How was governmental science best and properly controlled—through a democratically political or a politically elitist mechanism?" (Kevles 1987: 54–55). Whether research policy should be centralized in a single agency or carried on through a network of distinct agencies was also highlighted by the report.

The period from the Civil War to the onset of World War II answered the latter question by default. During the period from 1900 to 1940, more than forty science-related agencies were created by the federal government (Redmond 1968: 174). The scientific bureaus and laboratories created certainly up until the 1930s—including the National Institute of Health, the National Bureau of Mines, and the National Bureau of Standards—typically had a problem orientation, providing routine services in line with mission of the government agency or playing a regulatory role (Geiger 1986: 60; Kevles 1988a: 116, 118). And with the exception of agriculture, the government's role in academic scientific research was minor prior to World War II (Kohler 1987: 135).

World War I provides major exceptions to all of these general observations in the period between the Civil War and World War II, as it prompted a major institutional crisis and led to institutional reorganization, which set several precedents of relevance to the post–World War II period. An agency created to aid the war effort in World War I, and one that had decisive influence on the organization of science during World War II, was the National Advisory Committee for Aeronautics (NACA). Founded in March 1915 to supervise and coordinate American aeronautical research, the committee included representatives from the armed forces and several other government agencies, "but the chairman and the majority control reside[d] in a body of citizen scientists appointed by the president who, in practice, . . . followed the recommendations of the chairman in appointments to fill vacancies" (Compton 1943: 74).[12]

NACA was slow getting starting during World War I. Its initial appropriation was only $5,000, and at the war's end the committee's first wind tunnel was not yet complete (Dupree 1957: 318; Kevles 1987: 105). Yet, in promoting aeronautical research, the committee let research contracts to industrial firms and universities (Kevles 1987: 293), and NACA's domination by civilian science and its contracting procedures provided important models for Vannevar Bush's Office of Scientific Research and Development (OSRD). Bush himself served on NACA in 1938 and was appointed chair in 1939 (Kevles 1987: 246). Beyond giving Bush an opportunity to see NACA's administrative practices close up, this appointment allowed Bush to establish important social contacts in the federal government.

The Naval Consulting Board (NCB), created in 1915, was a second World War I agency created to centralize research organization. The board, under the chairmanship of Thomas Edison, was dominated by the engineering profession and devoted most of its energies to reviewing proposals submitted by civilian inventors to aid the war effort (B.L.R. Smith 1990: 29). Few of these were taken up, and the navy ignored all of the devices proposed by Edison himself (Kevles 1987: 138).

In addition to NACA and NCB, as I have noted, World War I saw the creation of the National Research Council (NRC), a crucial mechanism for creating links between university and industry scientists, as well as between scientists and the federal government. An executive order signed in 1918, two years after creation of the organization, permitted the NRC to "promote research in the name of the government yet to remain a private organization beyond government control" (Kevles 1987: 140).

The NRC supervised important work in such areas as submarine detection (Pursell 1966: 236). But while the council did serve as a clearinghouse for information and a focus for scientific personnel, according to Dupree, the organization "never developed an adequate full-time administration to direct all phases of its program as a unit" (1957: 323). In any case, the efforts at coordination and central administration of research policy by NCB and the NRC were quickly reversed after World War I, and "the Federal Government resumed its support of mission-oriented research through the separate programs of its many agencies and departments" (U.S. House Task Force on Science Policy 1986a: 11).

If these efforts were in some sense fleeting phenomena, they each established important models and precedents for the organization of research during World War II and afterward. They established the appropriateness of scientist control of research policy. NACA established the contract method of supporting research, and the NRC helped legitimize the concentration of federal research resources in those institutions that already dominated university research. In addition, the NRC in particular highlighted the issue of the central coordination of research policymaking.

Conclusion

In this chapter, I have attempted to create a portrait of the scientific field in the twentieth-century United States as a complex configuration, in part constituted by, and in part overlapping with, what might be broadly termed the economic and political fields. It is an entity cross-cut by universities, foundations, industry, and the government. This is not to say the field does not change. It is as much as anything a historical phenomena and, as I have indicated, the period from about the Civil War to the onset of World War II saw important changes in several components of the scientific field and in their relationship to one another. The turn of the century marked the rise of the research university and of industrial research, as well as the development of foundation funding as an important mechanism for research support. World War I linked the various components of the scientific field in a way they had not been linked prior to that time.

The period beginning around the turn of the century saw the slow rise in the importance of scientific research—the expansion of scientific authority—in all areas of American life. The seeds of elite scientists' collective advancement project took shape during this period. The need for vast resources for big science became evident. Furthermore, the payoff from scientific research also became clear, enhancing the cultural capital of the scientific elite. In addition, a range of precedents laid the groundwork for broad social acceptance of a central principle of elite scientists' project: autonomous control of science and of research funding. Finally, Vannevar Bush and his allies laid the basis during this period for important social connections inside and outside the government.

By World War II, there was no question but that scientists should play a prominent role in the war effort. Social connections made possible by the overlap of institutional spaces in the field, established especially by the National Research Council during World War I, constituted the network within which plans for research policy during World War II were developed, and these connections, and connections developed during World War II on the basis of them, provided a resource and cohesion for Vannevar Bush and his colleagues in their effort to define the federal postwar research policy agenda in the United States.

Importantly, this group drew on a prewar institutional legacy as their model for the wartime Office of Scientific Research and Development (OSRD). The OSRD was modeled—in its control by part-time civilian scientists and its contracting of research to universities and industry—on the National Advisory Committee for Aeronautics (NACA) of World War I fame.[13] The notion of contracting research, rather than undertaking the work in government laboratories, itself derived from the way in which foundations ultimately took to supporting research. Finally, the philosophy that shaped the perspective of Bush and his colleagues in their vision of a postwar research policy agency—its role in promoting "basic" as against "applied" research and in its policy on intellectual property—was shaped in important ways by the overlap between academic and industrial science.

3

A SCIENTISTS' WAR

Institutional Advantage, Social Connections, and Credibility

> The war was to an unprecedented degree a war of science.
>
> *Annual Report of the . . . Smithsonian Institution,* 1945

> The weapons whose evolution determined the course of war are dominantly the products of science, as is natural in an essentially scientific and technological age.
>
> Vannevar Bush, 1946

> The war effort taught us the power of adequately supported research for our comfort, our security, our prosperity.
>
> Vannevar Bush, 1970

When historians speak of World War II as a scientists' war or a physicists' war, they are typically referring to the tremendous technical achievements of science during this time and the importance of scientific research for the Allies' victory (Kevles 1987). Scientists associated with the United States' war effort were responsible for the development of radar, invention of the proximity fuse, the mass production of penicillin, and, most memorably, development of the atomic bomb.

But it was not just the technical achievements of scientists that made World War II a scientists' war. It was also the changed environment created by the war. Knowledge became power, and the dependence of the nation on scientists put a vanguard group of scientists, led by Vannevar

Bush, in the position to gain a powerful institutional space in the American state—the Office of Scientific Research and Development—and to use that location to enhance their social capital, their social connections, with civilian state managers and the military. This social capital combined with enhanced credibility or symbolic capital, which was a byproduct of scientists' role in the war effort, placed this scientific vanguard perfectly to shape the configuration of postwar research policy in the United States.

The crisis of war created a context in which institution building became possible.[1] With the onset of war, the state-building process leading to the genesis of the National Science Foundation began in earnest.[2] But, given the linkage between the war, the collective mobility project of U.S. scientists, and the ultimate contours of the federal research policymaking establishment in the postwar period, the account I provide in this chapter cannot be seen simply as a treatment of the initiation of state-building efforts; it must also be seen as a crucial stage in the collective advancement project of U.S. scientists.

A Scientific Vanguard and the Collective Advancement of Science

Magali Sarfatti Larson argues that professions "attempt to negotiate the boundaries of an area in the social division of labor and establish their control over it" (1977: xii), and she suggests that professionalization is "an attempt to translate one order of scarce resources—special knowledge and skills—into another—social and economic rewards" (xvii). Professions seek to create protected markets in which knowledge is their market asset and in which they have a "privileged or exclusive right to speak *in* or *about* their domain" (Larson 1984: 35). Professions attempt to control this market through education or socialization more generally, and the state often serves as a sponsor or guarantor of the process (48).

The *collective mobility projects* of professional groups are part and parcel of a situation in which "expertise . . . increasingly provides a base for attaining and exercising power by the people who can claim special knowledge in matters that their society considers important" (Larson 1984: 28). Monopoly over expert knowledge, according to Larson, becomes the basis for social power and collective mobility. The projects are collective in the sense that it is only through "joint organizational effort" that professions are able to achieve their objectives of defining the role that characterizes their profession and retain monopoly control over the profession's credentialing system (Larson 1977: 67, 70).

Andrew Abbott takes a slightly different perspective on professionalization. He argues that professions are engaged in struggles with other professions over jurisdiction—professional boundaries—over the legitimate area of their expertise or the space in which they can claim the right to speak (1988: 2–3). There are important differences in the approaches to professionalization taken by Larson and Abbott. But both Larson's and Abbott's conceptualization and analysis suggest that professions seek power and resource enhancement on the basis of their knowledge monopolies.[3] Here, at this basic level, I argue that a scientific vanguard during and after World War II attempted to speak on behalf of the American scientific community and to advance the community's objective of assuring its collective autonomy and access to federal resources.

These scientists did not seek market power as against some other profession. Nor were they attempting to restrict control over the credentialing process. Instead, they sought to extend scientist control over federal resources for scientific research. In a sense, they were engaged in a jurisdictional dispute with members of the executive branch and with some members of Congress over whether their expertise entitled them to control federal resources for scientific research or whether decisions about the allocation of federal resources for science ought to be determined by some representative social body—by a group of nonexperts.

Scientists were well positioned to shape U.S. weapons research policy during World War II. During the second and third decades of the twentieth century, Albert Einstein became "science's first bona fide celebrity" (Kuznick 1987: 13). In the wake of Einstein's work on relativity, scientists came to be seen as "uniquely qualified to interpret the frontiers of knowledge" (254). And, as historian Peter Kuznick argues, scientists gained unprecedented prestige as a result of their apparent role in the prosperity of the 1920s (9–14). As Frederick Allen wrote in 1931: "The prestige of science was colossal [in the 1920s]. The man in the street and the woman in the kitchen, confronted on every hand with the new machines and devices which they owed to the laboratory, were ready to believe that science could accomplish almost anything" (Allen quoted in Kuznick 1987: 14).

As the war approached, the credibility or symbolic capital of science and scientists was high—in the sense of the prestige that resulted from popular perceptions of the results of scientific research. Within this context, with the onset of the war, there was a recognition that, as Bush aide and OSRD historian Irvin Stewart notes, while "previous efforts to bring civilian science into the program of weapon development were based on the theory that the Services would know what they needed and would ask the scientists to aid in its development," the potential created by science had overtaken the generals' imaginations (1948: 60):

> [Modern science had] progressed to the point where the military chieftains were not sufficiently acquainted with its possibilities to know for what they might ask with a reasonable expectation that it could be developed. The times called for a reversal of the situation, namely letting men who knew the latest advances in science become more familiar with the needs of the military in order that they might tell the military what was possible in science so that together they might assess what should be done. (6)

Scientists had a knowledge monopoly (Szelenyi and Martin 1988), and with a background of what was viewed as scientifically induced prosperity, scientists were in a position to translate the accumulated symbolic capital of individual scientists into "power outside of the [scientific] field of origin" (Larson 1984: 61).

Not all scientists were equally well placed to do this, however. As Bourdieu suggests, "as authorities, whose position in social space depends principally on the possession of cultural capital, a subordinate form of capital, [scientists] . . . are situated . . . on the side of the subordinate pole of the field of power" (1988: 36). Possession of large amounts of symbolic or cultural capital was not enough. The purest of the pure scientists simply lacked access to those with economic and political capital. It was a small group of scientific elites that was in a position to translate the symbolic capital of science into institutional capital—a place in the state. This group was able to funnel the broad symbolic capital of science through their own social capital—social connections—to gain access to, and ultimately control in, the state.

At the head of this elite vanguard was Vannevar Bush. Bush spent his early career primarily at MIT, moving fairly quickly into the administration of elite scientific research.[4] In 1916, Bush received his engineering Ph.D. jointly from Harvard and MIT. Three years later, he was appointed associate professor of power transmission in MIT's department of electrical engineering. By 1923, he was a full professor (Kevles 1987: 294).

Bush's achievements as a researcher were significant. He worked at the "border line of basic and applied research, applying the theories of Norbert Wiener to the construction of machines for mathematical analysis" (Greenberg 1967: 75). He was responsible for developing a sophisticated differential analyzer, which was, according to Daniel Kevles, "a remarkable mechanical antecedent of the electronic computer" (1987: 294). In addition, he did research on ballistics and also undertook investigations "in a more secret field" (Baxter 1946: 14). By 1935, according to Bush biographer Larry Owens, applications for the differential analyzer "reflected the frontiers of scientific investigation and included studies in atomic physics, astrophysics, cosmic rays, and seismology" (1987: 243–44). For his work on the differential analyzer and related instruments, Bush was honored with the Franklin Institute's Levy Medal in 1928 and received the Lamme Medal from the American Institute of Electrical Engineers in 1935 (*Current Biography* 1940: 13).

Despite his important technical accomplishments, Vannevar Bush is correctly remembered more as a science administrator than a scientist. In 1932, he was appointed dean of engineering and vice president at MIT (Chalkley 1951: 12), and he remained at MIT until 1939 when he became president of the Carnegie Institution in Washington (Dupree 1972: 449). While at MIT Bush worked closely with institute president Karl Compton in forming the Division of Industrial Cooperation and Research (Owens 1987: 289), and, by the "middle thirties, as Compton's right-hand man, [Bush] had become not only a major figure at the Institute, but a respected spokesman within the country's technical community" (vii). Even while at MIT, however, Bush was not simply a scholar or a scientist. He served as consultant to a number of companies, and in his early days at MIT he helped found the Raytheon Company, a New England electronics firm. Raytheon was not a major science-based company for many years after its founding, but the company grew "explosively in the 1950s on defense contracts" (Forman 1987: 160; Kevles 1987: 294) and by the 1970s was the largest employer in New England (Bush 1970: 168).

Also while at MIT, Bush became linked to the federal administration of science through his role on government boards. Bush was first appointed to the Science Advisory Board (SAB) headed by MIT's Karl Compton. The SAB was a Depression-era body created to provide science advice to government agencies and promote government support of scientific research (Auerbach 1965: 462–63). Later, in 1938 Bush was appointed to the National Advisory Committee for Aeronautics (NACA) (Baxter 1946: 14), which had been established in 1915 to supervise and coordinate American aeronautical research (Kevles 1987: 104–5).

When Bush moved to Washington to take over at the Carnegie Institution and, according to historian Richard Rhodes, "to position himself closer to the sources of government authority as war approached," he also became the chair of NACA (1986: 336). Indeed, "because the Carnegie was at the time the nation's largest private research organization outside the universities, the appointment [to the presidency of Carnegie] gave Bush entree into the highest levels of national scientific research and development policy" (U.S. House Task Force on Science Policy 1986a: 17).

While in Washington, Bush became connected to a wide range of corporations and scientific organizations. He sat on the boards of American Telephone and Telegraph (ATT) and Merck and Company, and he served as a regent of the Smithsonian Institution. He also served on the boards of such institutions as the Woods Hole Oceanographic Institute, Johns Hopkins University, MIT, the Research Corporation, and the Brookings Institution, and he served on the National Research Council's Committee on Policies.[5]

Bush was not unique in the type of social capital he amassed. A few select scientist-administrators found themselves similarly well positioned. Among them were MIT president and Bush-mentor Karl Compton, Harvard president James Conant, Johns Hopkins University president Isaiah Bowman, and National Academy of Sciences and Bell Telephone Laboratories president Frank Jewett. Karl Compton, a respected theoretical physicist, headed the government's Science Advisory Board (SAB), on which Bush and Bowman sat. All four men sat on the National Research Council's Committee on Policies. Bush and Compton sat together on the

Table 3.1 Organizational Interlock of the Science Vanguard, 1940s

	Organization[b]					
Individual[a]	ATT	Brookings	Hopkins	MIT	NRC	SAB
Isaiah Bowman	X		X		X	X
Vannevar Bush	X	X	X	X	X	X
Karl Compton		X		X	X	X
James Conant					X	
Frank Jewett	X				X	

Source: Data for this table are from the Papers of Vannevar Bush, Library of Congress.
[a]Bowman was president of Johns Hopkins University; Compton was president of MIT; Conant was president of Harvard University; Jewett was president of NAS and Bell Telephone Laboratories.
[b]ATT = American Telephone & Telegraph; Brookings = Brookings Institution; Hopkins = Johns Hopkins University; MIT = Massachusetts Institute of Technology; NRC = National Research Council; SAB = Science Advisory Board.

Brookings Institution's board,[6] and Bush, Jewett, and Bowman sat on ATT's board. Finally, Bush sat on the Hopkins board, when Bowman was president of the university (table 3.1).

These men were not only an elite vanguard at the pinnacle of the scientific community; they are also well characterized by Paul Hoch's suggestive term, "boundary elite" (1988: 87). Boundary elites can be thought of as groups of actors able to move and mediate between the elites of two or more spheres of institutional power. Among these spheres we would want to include the military, the (civilian components of the) state, and industry. Through institutionalized and informal contacts, Bush and his colleagues spanned the rather blurred boundaries between the military, the state, and industry.

Bush and his compatriots were not scientists in any simple sense, and it is the social capital they accumulated prior to the war that put them in a position to translate the symbolic power of science—the credibility of science—into a powerful position in the state. What is more, the multiply-constituted identities of these men shaped the philosophy that undergirded the OSRD and its predecessor agency, the NDRC. As well, they chose people to work in their agency with whom they were familiar from their travels in the prewar scientific field.

These men, and the elite factions of the scientific field they represented, truly saw scientific research and technology as saviors. In assessing the situation of the war, Bush remarked: "A highly technical war exists. The entire future may depend upon the skill and thoroughness with which the scientific and technical aspects of the war are developed by a part of the democratic organization of this country."[7] Bush and his colleagues further viewed scientific research as the linchpin of national economic and social welfare, and they clearly adhered to what might be termed an ideology of expertise. In Bush's view, science was fundamentally meritocratic, and he and his vanguard group feared the political control of research (Kevles 1977: 14–15; Reingold 1987). Their view is quite similar to the perspective of scientist-turned-philosopher Michael Polanyi. Polanyi argued that "the choice of subjects and the actual conduct of research is entirely the responsibility of the individual scientist, [and] the recognition of claims to discoveries is under the jurisdiction of scientific opinion expressed by scientists as a body" (1951: 53). In short, decision making about scientific research can only be undertaken by scientists—by experts.

Toward a Pivotal Role for Scientists: The Genesis of the National Defense Research Committee

The interlock established between members of the scientific vanguard led by Bush was important in early efforts to coordinate research for the war. As Bush tells the story and others confirm, the idea for the National Defense Research Committee originated in gatherings of the National Research Council's Committee on Scientific Aids to Learning. All four of the men whom I have described as the core of a vanguard of the scientific community were closely associated with the NRC, and all four, as well as California Institute of Technology Dean Richard Toleman, met regularly at New York's Century Club (Bush 1970: 32; England 1982: 4).

William Domhoff (1974: 92) has suggested that social clubs serve as important consensus-forming environments—environments within which informal policy is made—and it was at the Century Club that proposals for the NDRC first emerged. Historians have suggested that these men were concerned that, unless the government stepped in, certain important war materials might not be developed, and this would undermine any U.S. war effort (Penick et al. 1972: 60). In addition, according to one historian, Bush's own experience in World War I prompted his desire to improve liaison between military and civilian development of weapons (Rhodes 1986: 336). But it was not only a concern to improve government organization that prompted action by the Bush group. It was also a belief that the war would be a "highly technical" one and that, as Bush aide Irvin Stewart has noted, its prosecution could not be left in the hands of the military (1948: 6). The implication of this position is clear: organizing weapon development for the war could not be left in the hands of amateurs. Experts had to be in control, and Bush and his vanguard group were the experts. They knew—or at least believed they knew—best.

The model on which these men drew in developing their ideas for the NDRC was the National Advisory Committee on Aeronautics (NACA). Indeed, in developing his ideas for the NDRC, Bush relied heavily on advice from NACA's executive secretary, John Victory. According to Joel Genuth, "adopting the NACA model was the obvious way for Bush to make his method for building a research field the standard operating procedure of a new agency" (1988: 279). It was a way to institutionalize the ideology of expertise to which Bush and his colleagues adhered.

Given the contacts Bush had developed in Washington since his arrival, Bush's NRC committee colleagues left it to him to see to the organization's creation (Bush 1970: 34). Franklin D. Roosevelt confidante Harry Hopkins had earlier sought Bush's advice on a proposal to establish a National Inventors Council (Greenberg 1967: 76), and through his contacts with Hopkins and executive branch lawyer Oscar Cox, Bush arranged to meet with President Roosevelt in May 1940 (England 1982: 4; Kevles 1987: 297).

As Pierre Bourdieu notes: "Science gives those who hold it, a monopoly of the legitimate viewpoint" (1988: 28). Indeed, with the war looming, President Roosevelt recognized his dependence on science and scientists to win the war. Scientists' credibility had been on the rise since the Republican prosperity and the Einstein vogue. Edward Teller recalled Roosevelt saying at an early meeting that "if scientists in the free countries will not make weapons to defend the freedom of their countries, then freedom will be lost" (quoted in Rhodes 1986: 336). Thus, on June 27, 1940,

Roosevelt approved an executive order developed by Cox to create the NDRC, and "Hopkins assured Bush that the agency would have FDR's full support" (Kevles 1987: 297; see also Baxter 1946: 15; Bush 1970: 36).

The organization of the National Defense Research Committee (NDRC) set the stage for future wartime and postwar government science organizations. Similar in many ways to the National Advisory Committee on Aeronautics (NACA), the NDRC vested management in a committee of civilian scientists and liaison officers from interested agencies (Genuth 1988: 279). At Bush's suggestion, the NRC/Century Club group—including Jewett, Conant, Compton, and Toleman—was appointed to the committee. In addition, Commissioner of Patents Conway Coe and representatives from the armed forces sat on the committee (the original military members were Rear Admiral Harold Bowan and Brigadier General George Strong) (Baxter 1946: 15; Stewart 1948: 7). Roosevelt aide Cox was assigned to provide legal advice to the committee (Stewart 1948: 13).[8]

The method of using civilian scientists on a part-time basis allowed committee members to retain their full-time positions. Thus, Bush stayed at the Carnegie Institution, Conant at Harvard, and Compton at MIT (Stewart 1948: 26). In addition to these "volunteers," the NDRC included paid clerical staff and some paid scientists to supplement the work of Bush and his colleagues.

The committee's mandate required it to support scientific research on mechanisms and devices of warfare, except those related to flight (which was left to the National Advisory Committee on Aeronautics), and to supplement research done directly by the military (Stewart 1948: 8). To carry out its mandate, the leaders of the committee created five divisions, each headed by one member: armor and ordnance; bombs, fuels, gases, and chemical problems; communications and transportation; detection, controls, and instruments; and patents and inventions. Division heads were permitted to create as many subdivisions or sections as they deemed necessary, and this subdivisional administration was under the charge of academic scientists and representatives of several companies, including Bell Telephone Laboratories, Carbide and Carbon Chemical Corporation, Standard Oil Development Company, Ethyl Gasoline Corporation, Radio Corporation of America, United Fruit Company, and Jackson and Moreland (11–12).

Bush did not head a division, but had responsibility for overseeing the organization as a whole and for serving as liaison with Congress, the president, and the military. This position provided him with an opportunity to build on his already formidable social capital. As chairman of the NDRC, Bush had direct access to the president, and, as Bush noted, "I knew that you couldn't get anything done in . . . [Washington] unless you organized under the wing of the President" (cited in Kevles 1987: 301).

The urgency of developing technology for the war meant that money was not hard to come by for the NDRC, and this significantly enhanced the committee's freedom. Roosevelt aide Harry Hopkins early on assured Bush that the committee would have "as much money as it needed from the President's emergency funds" (Kevles 1987: 297; see also Baxter 1946: 15; Bush 1970: 36), and the attachment of the NDRC to the Council of National Defense "gave Bush a line position in the government and with it the crucial ability to spend money" (Dupree 1972: 456).

To carry out its mandate, the committee decided that rather than promote the building of research facilities under its jurisdiction it would sponsor research done through agreement with existing government laboratories and on a contractual basis by universities (Genuth 1988: 280). In making these agreements, the committee was given complete latitude, including the freedom to promote projects that were neither requested nor desired by the military. This, of course, set a precedent for the ultimate project of Bush's scientific vanguard: to assure scientist control over government resources for research.

Although philosophically opposed to the centralization of research, Bush and his colleagues found it necessary to develop central research facilities at several universities, including the University of Illinois, the University of Chicago, Northwestern University, George Washington University, and the Carnegie Institution of Technology (Baxter 1946: 21). In addition, the distribution of resources by the committee was restricted to a relatively small group of companies and academic institutions (table 3.2). By June 1941, the NDRC had signed 207 contracts to forty-one universities and 22 companies.[9] Thus, in one year, the NDRC had become a major player in U.S. war preparations.

Creation of the National Defense Research Committee marked the beginning of the rise of a scientific vanguard in the United States. Bush and his colleagues drew on their social connections and credibility in the context of an emerging international crisis to create an organization that

Table 3.2 NDRC Contracts Signed with Academic Institutions as of June 1941

Institution	Number of Contracts
Brooklyn Polytechnic Institute	1
Brown University	1
California Institute of Technology	8
University of California	10
Carnegie Institute of Technology	3
Carnegie Institution of Washington	8
University of Chicago	9
College of the City of New York	1
Columbia University	5
Cornell University	1
Cornell University Medical College	1
University of Delaware	1
Drexel Institute of Technology	1
Franklin Institute of the State of Pennsylvania	2
Harvard University	13
University of Illinois	6
Iowa State College	4
Johns Hopkins University	3
Massachusetts Institute of Technology	20
University of Michigan	4
University of Minnesota	3
University of Missouri	1
National Academy of Sciences	3
University of Nebraska	1
University of New Mexico	1
Northwestern University	3
Ohio State University Research Foundation	3
Pennsylvania State College	5
University of Pennsylvania	3
Princeton University	10
Purdue Research Foundation	1
Rensselaer Polytechnic Institution	1
University of Rochester	2
Rockefeller Institute for Medical Research	1
University of Southern California	1
Stanford University	3
University of Virginia	2
Wesleyan University	1
University of Wisconsin	5
Woods Hole Oceanographic Institution	1
Yale University	2
Total	155

Source: "Report of the NDRC," June 28, 1941, pp. 56, 57. Roosevelt Papers, President's Secretary's File, Box 2, Folder: Safe: Bush, V.

gave them resources and direct access to the president while at the same time largely insulating them from Congress and other outside pressures. Under other circumstances, perhaps Bush and his colleagues would not have been given the autonomy and resources with which they were provided. But the administration in general, and FDR in particular, recognized their dependence on scientific research. As Roosevelt noted in reference to the atomic bomb less than two years after he approved creation of the NDRC: "Time is very much of the essence" (quoted in Rhodes 1986: 406). Bush and his colleagues also created new contacts through their organization, and their structural position as well as the precedents set by the organization proved central to the future course of research policy (Dupree 1972: 452).

From the NDRC to the OSRD: Rise of a Scientific Elite to Organizational Power

By spring 1941, the NDRC "was emerging as a pivotal organization in the military research and development complex" (Kevles 1987: 299). At the same time, the NDRC leadership viewed the current structure of the organization as inadequate for the development and implementation of research policy for the war effort. The NDRC lacked a role in development and initial procurement. In addition, there was no effective liaison between the NDRC and aeronautical development, and the organization had no role in medical research. What is more, Bush and others felt that, given the large sums of money needed to keep the organization operating, it would not be possible to rely indefinitely on the president's emergency fund; instead, it needed its own money directly from Congress (Bush 1970: 42; Stewart 1948: 36).

Bush forwarded a report to the Bureau of the Budget on the National Defense Research Committee's limitations, and in July 1941, a year after creation of the NDRC, President Roosevelt signed an executive order (prepared with the assistance of Bush and Conant) creating the Office of Scientific Research and Development (OSRD) as part of the Office for Emergency Management.[10] As Bush and his colleagues had hoped, the order gave this new organization a role in development as well as research, provided for a medical committee, and made the organization eligible for direct congressional appropriations. Bush was made OSRD chief, and the NDRC, with Conant as chair, was given an advisory role in OSRD (Kevles 1987: 299–300; Stewart 1948: 38). Scientists gained a kind of autonomy and power of which they could only have dreamt in the past.

The war provided Bush and his colleagues a rare opportunity: to expand—on behalf of a larger scientific elite—the political power of the scientific community.[11] In this time of crisis, they were able to convert subordinate forms of capital—symbolic and social—into an institutional space in the state. They defined the terms of their own success. The war dramatically increased the value of their capital, and they were able to reap larger and larger returns on their investment.

During the course of the war, the OSRD became a powerful agency. It controlled its own budget and was directly responsible to the president (Lasby 1966: 264). The organization developed a working partnership with the military and continued to use NDRC-style contracts, solidifying relations with industry and elite universities (Rowan 1985: 12–13). In addition, the OSRD developed the trust of Congress (Bush 1970: 134) and enhanced the positions of a range of scientists associated with elite universities.

As chairman of NDRC, Bush had direct access to and had begun to develop a relationship with President Roosevelt. In his role as OSRD director, Bush's relationship with Roosevelt became more intimate, and correspondence clearly indicates Roosevelt's respect for, and trust of, Bush. In a 1944 status report, for example, Bush "proposed dividing the work of development and ultimate production [of the atomic bomb] between the OSRD and the U.S. Army Corps of Engineers, bringing in the Army to build and run the factories as Bush had planned to do all along. Roosevelt initialed Bush's cover letter 'OK. FDR.' and returned it immediately" (Rhodes 1986: 412). Bush also became Roosevelt's informal science advisor: Roosevelt sought Bush's counsel on such vital matters as development of the atomic bomb, and Bush "operated at all times with the assurance of the President's support" (Stewart 1948: 50).[12]

The war and the urgent need for the development of military technology made the OSRD and Bush crucially important in the president's eyes. Bush and the president met regularly, and Roosevelt was eager to make sure that Bush's efforts proceeded smoothly. The president made certain that sufficient funds would be available for the work of OSRD.

On at least one occasion, Roosevelt used his influence to postpone congressional hearings at which Bush was expected to testify. In addition, with his own backing, Roosevelt passed along Bush's recommendations to the military.[13]

The success of the Office of Scientific Research and Development depended not only on the president's support but also on the military's. Although Bush was not universally respected and trusted by the military, a 1946 article in *Fortune* magazine asserts that gaining the confidence of the military "was perhaps Dr. Bush's main achievement as head of OSRD" (1946: 120).[14] Indeed, Bush had the basis for a solid relationship with the military. Besides working with military leaders on NACA (Stewart 1948: 49), during the war, Bush developed collegial relations with such central military figures as War Secretary Henry Stimson and Navy Secretary James Forrestal (Baxter 1946: 33).[15] He worked with Secretary Stimson on Manhattan Project matters, and Forrestal's respect for Bush was sufficiently high that he went to bat for the OSRD director when Bush thought he was out of favor with President Truman (Bush 1970: 133, 303). Then, at the war's end, military officials worked closely with Bush in developing postwar plans for OSRD and government-sponsored research.[16] In short, Bush's institutional position enabled him to expand his social connections, his social capital.

Beyond the relationship Bush developed with military leaders, the relationship between the OSRD and the military was institutionalized in several ways. Liaison occurred at multiple levels from project determination to field adaptation (Stewart 1948: 154). In April 1942, the Joint Committee on New Weapons and Equipment was established to undertake a strategic review of the weapons developed by the OSRD. Bush was selected as chair of the committee, and the committee was to report to the Joint Chiefs of Staff. The committee was charged with *coordinating* "the effort of civilian research agencies and the armed services in the development and production of new weapons and equipment" (Baxter 1946: 29).

Liaison between the OSRD and the military was further enhanced when discussions between Bush and high-ranking military officials in 1943 led to establishment of the New Developments Division of the War Department's Special Staff (Baxter 1946: 33). In addition, War and Navy Department representatives sat on an advisory council created in OSRD to advise and assist the director with "respect to the co-ordination of research activities carried on by private and governmental research groups as well as to facilitate the interchange of information and data between such groups and agencies" (Stewart 1948: 37, 46).

Of course, the success of OSRD demanded amiable relations not only with the president and the military, but also with Congress. During the course of the war, Bush and the OSRD gained the respect and trust of many in Congress. Bush was regularly in contact with congressional committees, and one historian has described him as an "impeccable witness" (Kevles 1987: 300). Bush himself bragged that the House Appropriations Committee came to trust and back OSRD. Of course, Bush had to keep Congress apprised of the OSRD's work and also met behind closed doors with congressional leaders to discuss atomic bomb developments (Bush 1970: 133). He said the experience gave him a confidence he had not had before "in the workings of the democratic process in time of war" (134).

And what of the relationship of scientists to OSRD? The war provided the basis for a power and prominence previously unknown by the scientific community. The OSRD operated in a manner in keeping with the dominant ideology of the scientists, and it was structured to protect the autonomy of scientists associated with it. Indeed, as Hunter Dupree argues, the "basic purpose of OSRD was to keep the exercise of scientific choice in the hands of scientists, who alone were [considered to be] in a position to judge the merits of a given line of research" (1972: 454). Certainly Bush believed in using his powerful position to insist on specific lines of research over the objections of the military and to refuse to entertain others (453). This approach is a clear indication of the ideology of expertise that guided Bush and his colleagues in their efforts at OSRD and in their collective project more generally.

Division chiefs from major universities and elsewhere were given "wide latitude in the formulation and execution of their respective programs subject to general supervision" (Stewart 1948: 78), and the research system, operating through existing government, industry, and especially university facilities, was highly decentralized. A conscious effort was made to allow scientists to continue to work at their home institutions: for example, a good deal of wartime radiation research was done at MIT's Radiation Laboratory established under NDRC contract, and important research on the proximity fuse was undertaken at the Cleveland laboratories of the National Carbon Company (Baxter 1946: 20).

Table 3.3 25 Principal Nonindustrial Contractors with the Office of Scientific Research and Development as of June 30, 1945 (ranked by total dollar value of contracts received)

1. Massachusetts Institute of Technology
2. California Institute of Technology
3. Harvard University
4. Columbia University
5. University of California
6. Johns Hopkins University
7. University of Chicago
8. George Washington University
9. Princeton University
10. National Academy of Sciences
11. Carnegie Institution of Washington
12. University of Pennsylvania
13. Northwestern University
14. Carnegie Institute of Technology
15. University of Michigan
16. Woods Hole Oceanographic Institute
17. University of Illinois
18. University of Iowa
19. Franklin Institute
20. Evans Memorial Hospital
21. University of Rochester
22. Duke University
23. Cornell University
24. University of New Mexico
25. Battell Memorial Institute

Source: Baxter 1946: 456.

Each key figure in OSRD appointed the people who worked under him on the assumption that "the top men in any field were known to each other" (McCune 1971: 48; see also Pursell 1971: 276). The same principle operated in the granting of contracts, leading contracts to be awarded disproportionately to institutions in the northeast and on the Pacific coast, as well as to "corporations usually represented among the OSRD leadership itself" (Stewart 1948: 58; see also Pursell 1971: 276). The largest amount of funds allocated to any university was granted to MIT, where Bush had been vice president and Compton president. Harvard, where OSRD's Conant was president, ranked third in the total amount of money OSRD provided to university contractors. And Bush's Carnegie Institution ended up in the top fifteen nonindustrial money getters (Baxter 1946: 456) (table 3.3). As historian Carroll Pursell put it, "in a great chain of personal and professional patronage, the four key members of OSRD . . . chose the heads of the various sections . . . from among their friends. These in turn appointed their friends to head subcommittees, and so on down the line" (1979b: 368). The institutional space they had captured gave Bush and his colleagues an opportunity to extend resources and control to a tight network of colleagues and to begin to realize their collective advancement project.

The Technological Accomplishments of OSRD and the Status of Science

An account of the Office of Scientific Research and Development would not be complete without a discussion of the technological accomplishments of the agency and of scientific research in general during the war. This is because the position of the science vanguard was greatly enhanced by the technological achievements of OSRD and American scientists during the war. The organizational character of OSRD allowed Bush and his colleagues to extend and expand social connections that would facilitate the realization of Bush's National Science Foundation. But the technological achievements of scientists associated with the war effort provided Bush and his colleagues with a kind of credibility or symbolic capital that Bush's postwar opponents could never quite manage to surmount.

In the medical area, wartime saw the development of a number of drugs and techniques. OSRD research led to advances in blood substitutes and immune globulins to combat infections. Penicillin production was the most notable achievement of U.S.-sponsored medical research. Although discovered in 1929 by Sir Alexander Fleming and further developed by Howard Florey and Ernst Chain in the late 1930s, it was not until World War II that production levels of penicillin became sufficient for the drug's use in widespread treatment of human infection. The antibiotic was central to saving lives on the battlefront. Wartime research also brought malaria under control. A 1946 *Fortune* magazine article went so far as to suggest that military-related medical research during the war was the "spearhead in bringing the Army's death rate from disease down from 14.1 per thousand in the last war [World War I] to 0.6 in this" (1946: 117; see also Bush 1960 [1945]: 10, 49, 53).

Advances in weaponry and other equipment were also central to the war effort. Of course, the most famous accomplishment in weaponry is the atomic bomb. The possibility of constructing an atomic device was made evident by research done under OSRD auspices. When it was determined that developing an atomic bomb would be extremely costly, the project was transferred to the newly created Manhattan District of the Army Corps of Engineers. While the transfer eliminated direct OSRD oversight in the project, Bush and his colleagues retained an intimate advisory role in the effort, and, indeed, Bush worked directly with Major General Leslie Groves, chief of the Manhattan District, in his capacity as a member of the Scientific Advisory Committee to Groves and as a member of the Manhattan Project's Military Policy Committee (Stewart 1948: 49).

Besides the atomic bomb, radar developed during the war played an important role in the Allies' victory (Bush 1960 [1945]: 10). Radar permits the detection of objects beyond visual range; electromagnetic energy is transmitted toward objects, and the echos returned from them are observed. One of the most important microwave radar devices developed during World War II was a gunfire-control system that permitted anti-aircraft guns to fire accurately without searchlights or optics (*New Encyclopedia Britannica* 1991: 458, 462).

Another important technical achievement of wartime was the proximity fuse. Described as "one of the deadliest factors in U.S. mass artillery and anti-aircraft fire" during the war (*Fortune* 1946: 117), proximity fuse technology was developed at the Carnegie Institution and Johns Hopkins University under OSRD contract (Baxter 1946: 221–42). The device "was in effect a miniature radar in the nose of a shell to signal the proximity of any nearby object . . . and to detonate the shell at the distance calculated to do the most damage" (Dupree 1972: 455).

These achievements further increased the prestige of the scientific community. There was a kind of halo that enveloped the results of scientific research and scientists. Americans were especially awed by the atomic bomb. *Business Week* declared August 6, 1945, the day after the atomic bomb was dropped on Hiroshima, the beginning of the modern age (cited in Jones 1975: 119). According to Paul Forman, "the atomic bomb did not generate but mightily confirmed and reinforced a conviction, already firmly established in the minds of newspaper editors and other opinion leaders by the summer of 1945, that 'our national security rests upon superior science' " (1987: 156). As the *New York Times* noted:

> The greatest marvel is not the size of the enterprise, its secrecy, nor its cost, but the achievement of scientific brains in putting together infinitely com-

> plex pieces of knowledge held by many men in different fields of science into a workable plan. . . . The brain child of many minds came forth in physical shape and performed as it was supposed to do. . . . What has been done is the greatest achievement of organized science in history. (quoted in Boyer 1989: 185)

The Truman White House also saw the atomic bomb as the "greatest achievement of organized science in history" (cited in Kevles 1987: 334).

The successes of scientists during the war generated the widespread belief that scientific research was the key to progress, national welfare, and security. As Jones concludes on the basis of a review of newspaper accounts of the debate over wartime and postwar science policy:

> Americans expected science to give them great economic, medical, and strategic benefits. Science would facilitate the return to a peacetime economy by promoting new industries that would in turn create new markets and more jobs. It was believed that science would not only ensure the country's prosperity but would also, through scientific research in medicine, enable Americans to live healthier, longer lives. Finally, Americans expected science to provide the means to protect the healthy, prosperous nation from foreign aggressors. It was hoped that science would make the nation secure, at reduced cost and with minimal interference in the majority's pursuit of happiness. (1976: 39–40)[17]

The war indicated a need for more basic research and expanded scientific training (Lapp 1965: 4; Penick et al. 1972: 21), and the bomb dramatized the need for a national *coordinated* science policy (Jones 1975: 270–71, 274).[18] An editorial in *Colliers* in September 1945 commented that "it is hard to think of anything that could have dramatized more forcefully the need for a national scientific research policy, as advocated by Dr. Vannevar Bush and numerous others, than the loosing by the United States of the forces of the atom bomb on Japan" (quoted in Jones 1975: 270).

Conclusion

World War II marked an important turning point for American science, scientists, and science policy. An elite group of scientists became prominent spokesmen on science policy issues. Money flowed to scientific research to a previously unprecedented degree, and scientists gained a previously unknown public recognition.

In the specific case under discussion, the creation of the NDRC and later the OSRD marks the beginning of postwar state building for research policy. Important precedents or policy legacies were established that would play an important role in shaping the debate over postwar research policy. The government contract became an important means of supporting research. Indeed, between 1940 and 1944 the federal government went from doing most of its own research to contracting out for most of it (Penick et al. 1972: 100). The contract was important as a funding mechanism in itself and because its use established the principle that the government should fund research carried on by nongovernmental bodies, especially universities.

The division of research support was radically transformed. Before the war, two-thirds of research in the United States was supported by industry for commercial purposes. A little more than one-sixth was financed by the government, and slightly less than one-sixth was funded by universities and foundations. During the war, the federal government became the dominant sponsor of research, funding three-fourths of all research by fiscal year 1944 (U.S. Senate 1945/6: 10; U.S. Senate 1945a).

A group of elite scientists mostly associated with the physical sciences successfully administered the nation's massive wartime research program. In doing so, they established the principle that science should be governed by scientists, and this issue became central to postwar policy debates. In addition, an alliance was solidified between scientists, industry, and the military that would influence postwar research policy. It is perhaps not a coincidence that representatives of many of OSRD's major industrial contractors had a hand in shaping postwar research policy. Bush and others learned to work with and gained the trust and respect of the military; indeed, Bush's membership on a wide range of military-related policy committees virtually made him a member of the military elite.

The structural position of a small group of elite scientists was greatly enhanced during the war. Drawing on a powerful social network, Bush and his colleagues transformed their social and cultural capital into a powerful institutional space in the federal government. In turn, this location provided the opportunity to strengthen social connections and enhance the credibility of scientific research, and, in turn, these connections and this credibility provided Bush and his colleagues with a distinct advantage in their efforts to define the contours of postwar research policy.

In the history of the genesis of the postwar federal system for research policymaking in the United States, state building and the collective advancement of scientists are inextricably linked. The war provided scientists the opportunity to transform their knowledge monopoly into power: to gain resources and control. The expansion of the state—the creation of a scientist-*controlled* federal agency—was a particular aim of the Bush-led science vanguard. The war—the kind of crisis that makes institutional change possible—improved the general prospects for the creation of a single postwar research policymaking agency, and the achievements of scientists and the successful institutional work of Bush and his colleagues positioned the Bush-led science vanguard perfectly to shape the configuration of the federal system for research policymaking in the postwar period and to simultaneously realize the collective interests of scientists in autonomous control over massive resources for research.

4

HIGH HOPES

Setting the Agenda in the Battle for a Postwar Research Policy

---

Science is offered a choice—Freedom as outlined in *Science—the Endless Frontier* or regimentation.

John Teeter, Aide to Vannevar Bush, August 1947

Both national security and the betterment of living call for a constructive governmental policy toward scientific research.

Senator Harley Kilgore, May 1940

Research needs to be coordinated carefully and the projects should be selected in terms of our national necessities, and not the accidental interests of various scientific groups.

New Republic, July 1945

World War II set the stage for the (re)construction of federal research policymaking in the United States. The war constituted a historic opening. Material circumstances changed. There was widespread agreement that existing institutional arrangements could neither adequately facilitate prosecution of the war effort nor promote national well-being after the war. Indeed, the war generated broad agreement on the importance of scientific research and technology for the welfare and security of the nation.[1] In addition, consensus on the need for some form of national research policy emerged from the war (Bronk 1975: 409; Jones 1975: 351; Kevles 1987: 334).

But if there was consensus on the need for change, there was no consensus on the nature of the necessary change. The initial struggle to define the contours of postwar federal research policymaking began well before the war ended. From the outset, the terrain of debate was defined primarily by two opposing camps. On the one hand, there was the Bush-led scientific vanguard and their scientist and business supporters throughout the country. On the other hand, there were advocates of a New Deal agenda for science led by Senator Harley Kilgore.

The scientists advocated a system of research policymaking that ceded control and resources to their number. They supported a program that emphasized basic research, from which they asserted economic benefits would obviously flow. In the interests of what they saw as national economic welfare, they advocated granting property rights to those undertaking the research, not the government who funded it.

Kilgore, by contrast, supported public control of federal research support. He advocated a program that would directly link scientific research with economic development, and he constantly asserted the public's right to material benefits from publicly funded research.

In line with the discursive turn in the human sciences, Cambrosio and his colleagues argue that "science policy practices [are] . . . first and foremost, *representational practices*" (1990: 196). While I do not believe that Cambrosio and his associates provide a convincing case for the "first and foremost" portion of their claim (see Kleinman 1991), they are certainly correct that understanding the realm of discourse and representational practices is fundamental to understanding social practices. Surely any account of the establishment of a postwar federal agency for making research policy must consider the discursive character of struggle and debate.

To understand early efforts to define the terms of debate in the struggle over postwar research policy, one must look at the intersection of available discursive legacies, institutional structure, the institutional location of actors, and the corresponding range of capitals available to actors. In this chapter, I explore the early efforts of Kilgore and his colleagues on the one hand and of Bush and his colleagues on the other. Ultimately, the way in which the positions of the two sides became ossified in the early days of the debate continued to reverberate throughout the struggle.

Harley Kilgore and a New Deal for Research

Any historical outcome is the product of a range of broadly structural factors that define a range of opportunities and limits—factors including institutions, policy legacies, and discourses—and conjunctural or even highly contingent factors. To understand the early development of the agenda championed by Harley Kilgore, we must look at the intersection of a range of such factors, including Kilgore's own biography, his rise to institutional power, the discursive legacy of the New Deal on which he drew, and the discourse of science that he had to confront.

After a split in the West Virginia Democratic party in 1940, Harley Kilgore, a criminal court judge in West Virginia active in Democratic politics, was selected by the party to run for the United States Senate (Maddox 1979: 21–22). Elected on Franklin Roosevelt's coattails, Kilgore was a true New Dealer with a distrust of monopolies that dated from the days when his father was driven out of business by Standard Oil (Kevles 1987: 343; Maddox 1979: 26).

In his campaign, Kilgore expressed concern for resource mobilization in the event of war, and he called for drafting men and industry for the war effort (Maddox 1981: 14). Upon assuming office, Kilgore became a member of the Truman Committee—the Special Committee to Investigate the National Defense Program. On the committee, Kilgore saw evidence that the government was failing to manage preparations for war efficiently and came to view this failing as a result of overlapping bureaucracies and inadequate planning (Chalkley 1951: 7; Maddox 1981: 52, 78). With some in the Roosevelt administration, he became outraged at the abuse of patents to restrict trade and the flow of information (Leuchtenburg 1963: 259).

In hearings on technological mobilization, Kilgore made his concerns clear: "It is because I personally have seen in our war effort so much of these technological inadequacies and scientific defects—so many patents frozen and new inventions blocked, so little attempt to develop the initiative of inventors—that I have become intensely interested in the need for adequate and total technological mobilization" (U.S. Senate 1942: 6). He was concerned about the specter of resource shortages, like the shortage in rubber. Outraged, he reflected: "To a man from Mars it would seem incredible that we who possessed the greatest mobility on wheels of any nation on earth should have neglected as we had the technology of synthetic rubber—since rubber was the product upon which our entire mobility depended" (5). The government, Kilgore contended, lacked a "coordinating agency" to assess such problems and seek solutions, and business "disputed the best methods" in terms of their own interests (5). He argued that putting so-called dollar-a-year men—businessmen paid a dollar a year to serve part-time on government boards—in decision-making positions in the federal government created a political environment where those in charge were not accountable. This, for Kilgore, was intolerable (Kevles 1987: 344).

In short, Kilgore had three basic concerns. First, he was troubled by the lack of *coordination* between federal government agencies themselves and between federal government agencies and private firms, nonprofit organizations, and universities in promoting science and technology research. He believed that this organizational "chaos" would lead to the failure to develop technologies essential for the nation's economic and social welfare. Second, Kilgore was distressed about the *concentration* of research resources in a very few universities and firms. This concentration upset Kilgore's notion of fairness, and he believed concentration might lead to restrictions in accessibility to the products of research. Finally, he was concerned about *restrictions on the flow of research results* that should be available for public benefit. Private rights to patents of government supported research, he believed, were not conducive to the wide circulation and use of research results, and, of course, industry had a systemic motive—profit—to restrict access to research results, even if firms had no use for the results.

Driven by these issues, between 1942 and 1949 Kilgore proposed several plans for reorganizing the federal role in scientific research. While Kilgore's proposals varied in detail, they all suggested the same philosophical underpinnings. First, Kilgore believed that the "chaos" in research support as well as the circulation of research results demanded the *centralization* of research policy in a single government agency. This agency would *coordinate* all government research-related activities and, with the private sector, coordinate research activities nationwide. Second, Kilgore believed that agenda setting for research could not be left up to some "invisible hand" in the scientific community or in industry. Instead, his proposed agency would *establish research priorities* and make certain they were carried out. Third, Kilgore's agency would not simply focus on basic research; it would promote *the transformation of basic research into publicly usable results.*[2] Fourth, he believed that control of federal research dollars as well as decision making for the nation's science agenda more generally should be in the hands of a *broadly representative* government body. In its early guises Kilgore's plan called for a board with representatives from a wide array of social interests. Later, bowing to political pressure, he fought simply to maintain *public accountability* by requiring that board members and the director of the federal science agency be selected by the president. Finally, Kilgore's agency would promote an *equitable distribution* of federal research resources and the *broadest possible circulation* of the results of federally sponsored research.

The background of domestic crisis generated by the war viewed through New Deal lenses sparked Kilgore's interest. Appointment to head the Subcommittee on War Mobilization of the Senate Committee on Military Affairs gave the West Virginia Senator an *institutional soapbox*—the legitimacy and resources—from which to pursue his concerns. The subcommittee was well funded (*Fortune* 1946: 212), and according to his biographer, Robert Maddox, Kilgore became an innovator in his use of the congressional committee system (1981: 53). He gathered around him a group of young innovative thinkers, establishing something of a mini–brain trust.

Kilgore worked especially closely with Herbert Schimmel, a physics Ph.D., who had worked for the Works Projects Administration, as well as the House Committee on Defense Migration and the Senate Small Business Committee. In these earlier appointments, Schimmel studied problems in the American economy, and he was involved in shaping the new wartime agencies (Chalkley 1951: 5–7; Maddox 1981: 52).

Under Schimmel's leadership, Kilgore's subcommittee pursued issues of scientific and technological mobilization. Schimmel was behind the introduction of Kilgore's first major research policy proposal, his Technology Mobilization Act (S.2721). Like Kilgore, having studied the rubber crisis closely, Schimmel believed that the federal government would be better able than industry to undertake important research on and development of critically needed materials. He believed that monopolistic industries had no incentive to develop new and innovative products and techniques (Bronk 1975: 409; Hodes 1982: 107).

Kilgore's Technology Mobilization Act was introduced on August 17, 1942, with the approval of Senators Claude Pepper and Harry Truman (Bronk 1975: 410; Maddox 1979: 23). In promoting his position, Kilgore argued that "the research policy of the United States Government has been utterly chaotic ever since we founded the Republic, and I want to see it get back to a sensible foundation" (Kilgore quoted in Maddox 1981: 171). Kilgore hoped his legislation would guarantee victory in the war by mobilizing technical personnel, equipment, and public and private facilities; by breaking bottlenecks in the flow of material resources, which inhibited production; and by maximizing technical mobilization by making patents and all technical knowledge available for the war effort (U.S. Senate 1942: 1).

To achieve the objectives specified in the bill, the legislation called for creation of a new government Office of Technical Mobilization. The proposed office would be directly responsible to the president and would collect information on the nation's technological resources. It would review established production facilities, techniques, and products. In addition, it would have the right to draft personnel necessary for the United States to achieve technological preeminence in the war effort. Beyond its drafting capacity, the bill permitted the office to erect pilot plants and establish research facilities to fulfill its responsibilities. Finally, the bill gave the office the right to *selectively* promote the development of certain technologies through grants and loans and the right to patent and license products the research for which was supported by the Office (U.S. Senate 1942: 2–3).[3]

Hearings on the bill were held in October 1942, and some fifty-five witnesses were called (U.S. Senate 1942). The hearings drew attention to the inefficient use of scientific resources, as well as the concentration of government R&D contracts and policies that might lead to what Kilgore saw as government giveaways of patent rights (Kevles 1977: 10). In terms of resource utilization, one witness, Lyman Chalkley of the federal Board on Economic Warfare, reinforced Kilgore's concerns, arguing that "we have a great foundation of scientific fact which hasn't grown up and been used in our technology; and in this wartime," he concluded, "we have to develop technologies that are not profitable" (U.S. Senate 1942: 16). The implication here is that the government must step in to develop technologies that industry will not.

Chalkley also claimed that there were too many government agencies doing war-related technology work and pointed to the *centralized character* of such work in Britain as a superior way in which to organize technology development for the war (U.S. Senate 1942: 9–10). In line with the Kilgore proposal, other witnesses also recommended "amalgamation" of government functions to avoid duplication (72). Still other witnesses, like Hiram Sheridan, a self-proclaimed amateur inventor, pointed to the inadequate use of human resources in the war effort. Sheridan suggested that it was difficult for individual inventors to bring their inventions to the attention of the public (153).

In his testimony, Chair of the War Production Board Donald Nelson pointed to work on the electrolytic application of tin as part of the war effort. He suggested that this development combined the "work of a lot of people," and Kilgore pointed out that the development was made "at the expense of the Government" and suggested that, given the context of its development, a "private monopoly" should not be permitted in such products (U.S. Senate 1942: 280). In reflecting on this type of situation, Kilgore commented that "it seems to me when the taxpayers of the United States pay for the development of something, it is a crying shame to make them dig down in their pockets and pay a big royalty to some outfit that has grabbed off the results of their research" (quoted in Maddox 1981: 171).

The largest prewar government contracts went to firms and nonindustrial contractors that already had the greatest research laboratories (Kevles 1977: 6–7; McCune 1971: 48). The concerns of Kilgore and others about the concentration of government contracts seemed to be borne out by the war years. A study done in 1945 by Kilgore's subcommittee found that the concentration of government contracts was dramatic: "Of about 200 educational institutions receiving a total of $235,000,000 in research contracts from the Government, 19 universities and institutes accounted for three-fourths of the total. Of nearly 2,000 industrial organizations receiving a total of almost one billion dollars in research contracts from the Government, less than 100 firms accounted for more than half of the total" (U.S. Senate 1945a: 39).

There were supporters of Kilgore's early efforts, including several people who testified at Kilgore's 1942 hearings. In testimony before Kilgore's subcommittee and in an article in the *American Mercury* entitled "The Case for Planned Research," Waldemar Kaempffert, a science reporter for the *New York Times,* argued that "Senator Kilgore is merely trying to bring order out of chaos" (1943: 442). In supporting Kilgore's efforts, Kaempffert concluded that "*laissez-faire* has been abandoned as an economic principle; it should also be abandoned, at least as a matter of government policy in science" (445). He suggested that *planning* technological development was necessary to make sure development occurred to advance the nation's social and economic welfare (U.S. Senate 1942: 67, 69, 71).

Kaempffert, as vocal a supporter of Kilgore as he was, did not represent an organized constituency, but Kilgore did have some organized support outside the Congress. The American Association of Scientific Workers (AAScW) worked closely with Kilgore's staff and lobbied behind the scenes (Hodes 1982: 115–16). But the AAScW was not a particularly influential organization. Although a number of prominent scientists belonged to the group early on, it was dominated by biological scientists, and it was the physicists and chemists who were associated with the war effort and had credibility and connections (175). A related reason for the organization's relative lack of influence was that it lacked ties to Bush's Office of Scientific Research and Development (80). Interestingly, Bush's colleague Karl Compton was the organization's first president, but, according to Elizabeth Hodes, who has undertaken a detailed study of the AAScW, Compton "seems largely to have been a figurehead . . . , and [in any case,] he . . . was only superficially involved with the day-to-day activities of the AAScW" (65–66). In addition, organization members lacked political experience, the association was highly decentralized, and, in the end, it was plagued by red baiting.[4]

In the Roosevelt administration, too, Kilgore had support. Donald Nelson, Chair of the War Production Board, gave generally positive testimony at Kilgore's hearings. He argued that the United States needed to make civilian and military production work together. He pointed to the current misallocation of both material resources and personnel. And, while he generally supported the idea of a single central office to coordinate technology-oriented research, he expressed concern about the office's possible interference with Bush's Office of Scientific Research and Development and with the military and scientists more generally (U.S. Senate 1942: 273–88).

From the Justice Department, Special Assistant to the Attorney General Robert Hunter expressed concern in testimony about the use of patents by companies to suppress important inventions. He argued in favor of government control of patents resulting from government-sponsored research and concluded that "I personally feel that this bill is a piece of vitally needed legislation" (U.S. Senate 1942: 481).

But the Roosevelt administration was not united in its support for the legislation. Importantly, the military opposed Kilgore's proposal. Major C. C. Williams, a War Department member of the National Defense Research Committee, considered the Kilgore plan an overly ambitious attempt during wartime to make a comprehensive assessment of technological manpower and facilities. He argued that the National Inventors Council was already attempting to mobilize technological personnel, and he suggested that there was no reason to believe Kilgore's Technology Mobilization Office would do a better job. Suggesting perhaps a fear that the Kilgore office would encroach on the military's turf, Williams concluded that "the present organizations [including the OSRD and agencies within the military] do now provide for and have efficiently mobilized the scientific and technological personnel of the country" (U.S. Senate 1942: 658).

None of the major figures associated with the OSRD testified against the bill. Informally, however, Frank Jewett, a friend and colleague of Bush's and a member of what I have termed the scientific vanguard, expressed his opposition to the bill and made quiet efforts to stop it (Pursell 1979b).[5] His position was almost the mirrored opposite of Waldemar Kaempffert, the *New York Times* reporter:

> The basic arguments against the bill in the fields of science and technology are identically the same as those against corresponding bills in the fields of labor and education, mainly that it would completely revolutionize our entire American age-old concept and by placing vast powers in the hands of federal bureaucracy would set the stage for the complete domination of the life of the nation by a small group of federal officers and bureaucrats. (Jewett quoted in Pursell 1979b: 374)[6]

Jewett was openly conservative politically. But if there is a subtext to Jewett's position, it reflects a discourse of science suggesting that scientists must control decision making concerning scientific research and the resources for that research. Civil servants are not qualified to make decisions concerning science.

Kilgore feared the impact opposition among scientists would have on the prospects for his proposal, and he attempted to work with Jewett (Hodes 1982: 109).[7] In the end, however, the bill did not move beyond Kilgore's subcommittee, and, recognizing the political force of scientists' objections, Kilgore permitted his bill to die quietly at the end of the 77th Congress (Chalkley 1951: 10).

In light of objections to his previous bill and fear of scientific opposition to his legislation, Kilgore introduced a modified piece of legislation—his Science Mobilization bill, S. 702—on March 5, 1943, early in the first session of the 78th Congress. Kilgore made an effort to make the objectives of the bill somewhat broader than its predecessor and to eliminate some of the features of the previous bill that opponents felt were arbitrary and dictatorial (Bronk 1975: 410; Chalkley 1951: 10).

While Kilgore's first bill evoked little public reaction, the new one created more of a stir. It mandated the establishment of an Office of Scientific and Technical Mobilization to "mobilize the scientific and technical resources of the Nation" (*Science* 1943: 407). Much like Kilgore's earlier legislation, S. 702 called for an assessment of the current level of utilization of scientific and technical knowledge, along with facilities and personnel, and the development of policies for maximizing utilization of these resources in the national interest. While the earlier bill called for the selective promotion of certain technologies, S. 702 aspired to create a more comprehensive basis for a protoindustrial policy, giving the Office of Scientific and Technical Mobilization the right not only to promote technological development but also to do this within the framework of promoting full employment and providing assistance to industry and agriculture.

Kilgore's science and technology mobilization legislation, moreover, was not only directed at enhancing the war effort. The bill aimed to create *a single centralized agency* that would promote economic advancement in the postwar period through support for scientific research and technology development. The office envisioned in the legislation was intended to serve as the primary government advisor on science and technology issues. It would promote small business by making the benefits of scientific advance freely available to that sector of the economy, and it would undertake studies on the impact of new research developments and new technologies on America's national welfare. Finally, like its predecessor legislation, S. 702 authorized the office it created to acquire patents for inventions developed with government support and to grant nonexclusive licenses for their use.

Like Kilgore's earlier legislation, this bill designated a director or administrator appointed by the president to operate the proposed agency. Unlike the earlier bill, however, this piece of legislation proposed creation of a *full-time board* chaired by the director of the agency. The board—the responsibilities of which were left to the agency director—was to include six representatives in addition to the director, including a representative each from industry, agriculture, labor, and the public, as well as two scientists or technologists. In addition to the board, the act proposed creation of a larger advisory committee, including the board members as well as representatives from government departments appointed by the president, additional representatives from business, labor, the "consuming public," and scientists and technology professionals (*Science* 1943: 407–12).

Kilgore's efforts fit squarely into a New Deal tradition. They were part of a New Deal discourse of planning, coordination, social cooperation, and democracy. Although there was division within the Roosevelt administration, there was a substantial contingent who advocated *social and economic planning*. Moreover, during the first Hundred Days, the Roosevelt administration "committed the country to an unprecedented program of government-industry cooperation . . . [and] agreed to engage in far-reaching regional planning" (Leuchtenburg 1963: 61). The Tennessee Valley Authority amounted to an experiment in social planning, and the National Resources Planning Board was an effort to increase *coordination* of national social and economic policy (Leuchtenburg 1963: 54–55; Weir 1988: 165). Kilgore's effort to establish a centralized science and technology agency to coordinate policy, eliminate duplication, and establish priorities in the interest of the nation's social welfare fits directly into this tradition.

Kilgore's proposals are also consistent with the executive reorganization plan made by President Roosevelt after his 1936 election. The plan aimed to facilitate *coordinated* state intervention in the economy and into society and called for consolidation of the federal administration into *centrally controlled* cabinet departments (Skocpol 1980: 193). The reorganization plan failed, and in the end, opposition in Congress to what was perceived as an executive power grab may have hurt Kilgore in his later efforts to create a National Science Foundation with representatives appointed by, and accountable to, the president.

Kilgore's attempt to give substantial authority in governing his proposed agency to bodies with representatives from a broad cross-section of social interests also had precedent in New Deal policy initiatives. Indeed, "by the end of the thirties, the New Deal had extended the idea of popular involvement to a host of different agencies" (Leuchtenburg 1963: 86). For example, operators and miners were included in setting coal production quotas; similarly, local advisory committees were consulted on youth and arts projects (86), and, of course, the administration had a more general commitment to promoting business-government cooperation (35, 61).

Insiders suggest that Kilgore's staff spoke to potential hearing witnesses prior to the hearings on the senator's new bill and only permitted supporters of the legislation to testify formally.[8] Certainly the lopsided representation of pro-Kilgore witnesses at the hearings in the spring of 1943 supports this contention. Of the more than thirty witnesses who testified, none spoke against Kilgore's bill as a whole, and only one witness expressed reservations about specific provisions in the bill (U.S. Senate 1943: 136–41). No members of the Bush-led science vanguard and no representatives of scientific organizations testified on the bill.

A good deal of testimony was aimed at documenting abuses by business that led to restrictions in technological development. For example, Assistant Attorney General Wendell Berge described in detail the way in which "the Wisconsin Alumni Research Foundation [acted] . . . as a screen behind which a group of monopolistic chemical, pharmaceutical, and food companies control vitamin D." According to Berge, the Foundation "exhibited a lack of interest in research unless commercial advantage could be obtained . . . [and] attempted to suppress the publication of scientific research data" among other things (U.S. Senate 1943: 740, 741).

Representatives of small business and others testified in favor of the bill. Finley Tynes of the Staunton and Augusta County Virginia Chamber of Commerce declared that S. 702 "is desperately needed." He said he supported the bill because it would "make available to small companies research facilities and give them access to new inventions and processes to which they are now denied" (U.S. Senate 1943: 177–78). This view was echoed in testimony by United States Vice President Henry Wallace (703–11), who argued:

> The application of modern science should not be the exclusive domain of the great corporations and cartels who can, if they desire, restrict and suppress new inventions and scientific information to suit their own interests instead of the public interest. Unless the little man has access to the bounties of technology, free enterprise will suffer to the detriment of full employment of labor and resources. (U.S. Senate 1943: 705)

In addition to stacking the hearing witnesses, Kilgore attempted to head off potential opposition to his legislation among members of the scientific community. He tried to reassure scientists by indicating that they alone could "provide the advice and counsel to make the bill adequate to the national need" (Kilgore 1943: 152), and he implied that his bill did, indeed, have broad support among scientists and that it was only "vested interests" who opposed the bill (151).

During hearings, an image of corporate misconduct in preparation for war was created (Chalkley 1951: 12; U.S. Senate 1943/4). And in a letter that appeared in *Science* magazine, Kilgore expressed his fear that if a central science and technology agency was not created to, among other things, promote research in American universities, industry support of such research would continue to make university research "the handmaiden for corporate or industrial research" (Kilgore 1943: 152).

Despite Kilgore's control over the legislative agenda and efforts at image management, his bill still faced significant opposition. He was able to garner the support of some members of the scientific community, including once again the American Association of Scientific Workers (Hodes 1982: 111), but major scientific societies opposed the legislation (Bronk 1975: 410; Chalkley 1951: 12).[9] The executive council of the 500,000-strong American Association for the Advancement of Science (AAAS) opposed the legislation on the grounds that it was too sweeping, in many senses not necessary, and because, the organization concluded, the administration of Kilgore's Office of Scientific and Technological Mobilization would be "political" (American Association for the Advancement of Science 1943: 135–37).

While these were the explicit claims made in the Association's resolution against the Kilgore bill, a close reading of the resolution suggests that the organization's concerns reflected a belief that the Kilgore bill would take control of research and other aspects of the scientific enterprise from scientists. For example, the resolution contends the claim in S. 702 that scientific information in an "uncoordinated state" is inaccurate. The AAAS points to the fact that there are "scores" of scientific journals and thousands of abstracts published annually. It would seem, however, that this portrayal of the status of scientific information would support Kilgore's claim that the information is "unassembled" and "uncoordinated." Of course, coordination by an office of the federal government would mean taking the power to represent scientific knowledge partially away from the scientific community.

The AAAS's concern with the supposed political character of Kilgore's agency again speaks to the issue of the control of science by scientists. The AAAS resolution refers to the fact that to be considered a scientist under the Kilgore bill requires only six months of training. As Whitley has noted for science and Larson has noted for other professions, a central mechanism for protecting collective interests is through control of credentialing processes (Larson 1984: 48; Whitley 1984: 63, 82). This clause in the Kilgore bill clearly takes the monopoly of certification away from the scientific community. If that were not bad enough, representation on the board of Kilgore's proposed agency would include representatives of "nonscientific" interests. What is clear is that as genuine as the AAAS may have been in its opposition, its position strongly reflects a concern that passage of Kilgore's bill would take control of science from scientists.

Other organizations of scientists, including the American Institute of Physics, also expressed opposition to the legislation (American Institute of Physics 1943). Perhaps most important, Bush and his elite colleagues opposed the legislation. Bush expressed his views in an open letter to Kilgore which appeared in the December 31, 1943, issue of *Science* magazine (Bush 1943). In the letter, Bush clearly laid out the tension between Kilgore's proposal to centralize the governance of research policy and the scientific community's need or desire for autonomy. Centralized control of science policy, Bush argued, makes sense during exceptional circumstances. The war, he believed, was such a circumstance, but his own Office of Scientific Research and Development was already doing the job, and he saw no need to create a new agency (572). After the war, Bush said, he would favor "a central scientific clearing house set up for the interchange of data and plans in order to prevent overlapping and to facilitate . . . cross-fertilization." But this agency would not control the research agendas of other government agencies, as Bush believed Kilgore's Office of Scientific and Technical Mobilization would (575).

Bush criticized Kilgore's bill for not giving sufficient control to scientists. He opposed the governing board proposed by Kilgore, suggesting instead that the United States needed "an advisory scientific group, drawn on a voluntary basis from the best qualified scientific personnel of the country." And determining who the best and most appropriate scientists were, Bush asserted, could only be done by scientists' colleagues (1943: 575). Bush said he favored government support of scientific research, but "without being attended by stifling control" because "science of the highest type flourishes best when it seeks its own objectives and pursues its own uninhibited ways" (577).

The rhetoric Bush used in his letter suggests that nonscientists have no role to play in governing science. The public can provide financial support for scientific research, but must stand on the sidelines while scientists seek to "increase the knowledge and understanding of man" (Bush 1943: 577). Whether intentional or not, such a discourse serves to legitimate the collective advancement of the scientific community. Scientists gain resources and control and are accountable only to themselves. As I noted in the previous chapter, this discourse predates the struggle over a postwar federal research policy agency. Nevertheless, it derived a good deal of power from the expansion of scientists' symbolic capital during and after the war. As we will see, this kind of rhetoric was used by Bush and his allies throughout the debate over a postwar research policy agency.[10]

Business interests had been relatively quiet on Kilgore's earlier bill, but they actively entered the debate on this one. Kilgore garnered some support from small business, but remained suspicious of "vested interests."[11] And his legislation was actively opposed by the National Association of

Manufacturers (NAM), who branded the bill "a comprehensive plan for the most ambitious project to socialize industrial research and technical resources that has ever been proposed in the United States Congress." NAM leaders were particularly adamant in their opposition to the bill's patent provisions, which they believed eliminated the "*incentive* given to inventors by the Constitution."[12] NAM focused its efforts on directing "public attention to the importance of keeping science and research free of government domination."[13] They sent a statement of their positions to scientific societies, prepared statements for public release, and requested that NAM members contact their legislators.[14] There is, however, no evidence of significant lobbying activities by NAM at this stage in the struggle.

Kilgore dropped his second bill, as he had his first. One historian suggests that Kilgore "appreciated many of the scientific community's dissatisfactions with the new measure" (Kevles 1977: 15), but in discussing support and opposition for his measure, Kilgore noted that letters to his office suggested "men of science favor the bill. The vested interests, and those who are influenced or controlled by the vested interests, are against it, and they are most unscientific in their attacks upon it" (Kilgore 1943: 151).

Kilgore was aware of the Bush-led science vanguard's opposition to the bill. He did not specifically exclude this group from his criticisms of what he termed "vested interests," and it seems likely he would have viewed critiques made by Bush, Compton, and others who focused on fear of dangerous expansion of bureaucracy as an "unscientific" screen for scientists' desire for control of the government's postwar science agency. Still, as I have noted, Kilgore was painfully aware of scientists' political power—their vast pool of symbolic capital and the vanguard's institutional and social capital. Thus, we should read Kilgore's "appreciation" of scientists' criticisms less as recognition of the correctness of their position than as cognizance of their power.

Kilgore had not given in. In 1945, his committee published a report that amounted to a detailed justification of his position. With the end of the war, the study suggested first that "the problem of national defense research [was] . . . inseparable from the reconversion of our research activity to high peacetime levels" (U.S. Senate 1945a: 3). It pointed to what it termed a "reconversion gap" in total U.S. spending on research by industry, foundations, universities, and government (a gap between pre-

and postwar spending levels) if there were not both public and private efforts to maintain wartime funding levels for research (11). Filling this gap would require, the report suggested, a substantial increase in federal support for research, though not to wartime levels (12). Support would need to be maintained for military and medical research as well as for basic research "upon which all applied science is built" (13).

In addition to establishing appropriate levels of research funding, the report pointed to the great number of federal agencies that had some role in research policy and to the temporary character of coordination between them that was arranged during the war. The report concluded that "some mechanism for *coordinating* Federal research activities must be worked out before we can arrive at an intelligent balanced use of federal research resources" (U.S. Senate 1945b: 13; emphasis added). Any coordinating mechanism, the report continued, should "insure that the most important problems receive adequate attention," match scientific problems with researchers and facilities, and promote the widespread distribution of research findings (13). The report recommended that final responsibility for such coordination be left in the hands of a *central scientific agency* or an overall supervisory agency such as the Bureau of the Budget. The federal agency would coordinate all federal and private activities through *cooperative mechanisms* and advisory machinery (13).

There would be a division of labor in allocating research resources, if the conclusions of the Kilgore committee report were followed. Applied research would largely be the task of industry. Basic research would be funded by nonprofit organizations and the federal government, since this work is not typically "immediately profitable" (U.S. Senate 1945b: 16). This division of labor posed coordination problems, but, the report concluded, "the problem of coordination, already difficult, would be greatly intensified if a number of new Federal agencies were to undertake to allot funds to universities" (14).

The report stressed the need for *well-planned* support in research priority areas and recommended the establishment of "a broad and representative board to advise on the use of university laboratories and . . . Government facilities in a well-planned program of basic research" (U.S. Senate 1945a: 16). Finally, the report expressed concern about the *concentration* of industrial research in a few large companies. A new government agency, the report suggested, could counter this by making important research results—especially those resulting from the war effort—available to small- and medium-sized firms (18). In a related matter, the report concluded that further discussion was needed on the disposition of patents resulting from government-funded research, but suggested that the results of federally funded research should be freely available (17).

From this report emerged a new bill, milder in tone, but similar in substance (see chapter five). Kilgore had defined his terms. He continued to support *a single centralized coordinating and planning agency.* He advocated support for basic *and* applied research to advance national social and economic welfare. He continued to express the need to promote the free flow of scientific information and the inappropriateness of private entities deriving exclusive benefit from government-supported research. Finally, he continued to promote *broad public involvement in decision making* concerning federal science and technology policy.

Kilgore drew on a powerful New Deal discursive legacy: a discourse of planning, coordination, social cooperation, and democracy. But New Deal programs did not constitute a solid and long-standing policy legacy. Unlike countries such as Britain where economic mobilization for the war meant the creation of permanent centralized agencies, in the United States, centralization, as a means of policy coordination was established only on a temporary basis. And as Amenta and Skocpol point out, "war spurred visionary social planning in America, but the political system allowed no bureaucratic continuity for, or moratorium on, social policy plans that had been devised during the war" (1988: 111–13). At the war's end, authority in postwar planning was appropriated by Congress, which quickly dismantled the temporary war-planning apparatus (Orloff 1988: 61).

The basis for the transformation of federal research policy Kilgore sought was a populist representation of a country gone awry. In speeches and in hearings, Kilgore and his supporters promoted an image of a government taken from the people and an economy controlled by "vested interests" for their narrow gain. This might have had a powerful resonance in the Depression, but on the heels of scientific success during World War II, a new discourse had captivated the public and elite actors alike. Kilgore faced a powerful discursive opposition—an ideology of science energized by scientists' success in the war, a promise of improved national well-being in exchange for scientists' empowerment. In addition to this powerful discourse, Kilgore faced a fragmented state with a disunified executive. What is more, his own political party could not enforce party discipline. Absent these factors, Kilgore might have scored a relatively easy and complete victory. But in the event, the struggle was long and drawn out, and Kilgore's victory was relatively hollow.

Science—The Endless Frontier: Bush and Best Science

While Kilgore's proposals rested on a less than secure policy/discursive legacy and faced opposition from certain segments of the business and scientific communities, as well as division within the executive, with an extensive staff, contacts in the executive and Congress and direct access to the president, the Office of Scientific Research and Development could hardly have been a better *institutional soapbox* from which to influence debate over the contours of federal research policy. The OSRD actively attempted to thwart the Kilgore subcommittee's progress. In 1944, OSRD staff looked into the Kilgore subcommittee's legislative authority and found that it did not extend to postwar research policy. Thereafter, they informed military scientific agencies that they need not cooperate with the Subcommittee on War Mobilization in its efforts to obtain information on the military's plans for postwar military research (Maddox 1981: 165).

The Office also pursued more direct efforts to influence the course of postwar research policy and counter Kilgore's initiative. Acting on a suggestion from Oscar Cox, an attorney in the Roosevelt administration and counsel to Vannevar Bush, Cox, Bush, and others on the OSRD staff drafted a letter intended for President Roosevelt's signature requesting from Bush a report on the role of the federal government in postwar research (England 1976: 41; Kevles 1987: 347; Rowan 1985: 41).[15]

Bush and his colleagues were successful in getting the president to issue the letter. Dated November 14, 1944, Roosevelt's letter praised the accomplishments of Bush's OSRD and indicated that the lessons of OSRD might profitably be used in the postwar period:

> The information, the techniques, and the research experience developed by the Office of Scientific Research and Development and by the thousands of

> scientists in the universities and in private industry, should be used in the days of peace ahead for the improvement of the national health, the creation of new enterprises, bringing new jobs, and the betterment of the national standard of living. (reproduced in Bush [1945] 1960: 3–4)

Toward this end, the letter asked Bush to answer four questions. First, the letter requested recommendations for rapid disclosure of wartime scientific advances consistent with national security. Second, it asked for recommendations on organizing a medical research program to fight disease. Third, it asked for recommendations concerning support for scientific education. Finally, and most important for the future of research policy, the letter requested advice on how the government should support public and private research in the postwar period.

To complete his report to the president, Bush appointed four committees, one to answer each of the questions.[16] Thus, there was a committee to look into science education, one to look into medical research, and so forth. And Bush and his committees were guaranteed whatever resources were necessary to complete the report.[17] Committee members received compensation, and the committees benefited from the assistance of the OSRD general counsel in preparing the report (Stewart 1948: 187).

Bush relied heavily on men with whom he had previous contact in establishing his committees. Several of the committee members had worked for OSRD. The committee established to address the question of government support for public and private research—the question of the most relevance to this book—was headed by Isaiah Bowman, president of Johns Hopkins University, where Bush sat on the board of trustees. In addition, Bush and Bowman were both members of the National Research Council and had worked together on the government's Science Advisory Board. The committee included scientists, academic administrators, high-level government bureaucrats, a foundation executive, and representatives of five companies. Six of the seventeen men who served on the committee had previously worked with OSRD in some capacity.

Among the companies represented on the Bowman committee was Bell Telephone Laboratories (BTL). BTL was represented by the company's research director, Oliver Buckley. Buckley was a member of the informal Directors of Industrial Research group (DIR), and, according to the group's records, he was "very anxious to secure the advice" of other research directors in the organization concerning the questions that the Bowman committee would be addressing. Although the records of the DIR do not indicate how the organization came down on the issue of government support for research, it seems safe to assume that they, like other industrial leaders at the time, took the position that industrial research must be based on fundamental research undertaken in the universities and supported by government and that government should not support applied research, which might compete directly with research undertaken by industry (Palmer 1948: 2042–44).[18]

The participation of Buckley and other representatives of industry *directly* on the Bowman committee provides one avenue through which industry expressed its interests and helped shape the agenda for the organization of U.S. research policy in the postwar period. Business interests also affected the agenda more *indirectly* through Bush, as well as Bowman. As I have noted, both men had intimate contact with industry. In the previous chapter I discussed Bush's long history of involvement with industry, as well as Bowman's links to the industry-dominated National Research Council. Bush, Bowman, and other members of the science vanguard were not scientists in any simple sense. As I suggested in the previous chapter, the science vanguard was really part of a boundary elite (Hoch 1988: 87), and while these men should not be seen as representatives of industry, their constitution at the line between science and industry undoubtedly shaped their perspectives on research policy. Indeed, in this context, it is not surprising that in his 1943 letter in *Science,* Bush opposed Kilgore's proposal for changes in federal patent policy. Kilgore's position was widely opposed by industry which believed that government acquisition of patents resulting from government-supported research and especially the granting of nonexclusive licenses for use of these inventions would undermine the economic incentive system.[19]

The report issued in July 1945 by Bush and his colleagues, entitled *Science—The Endless Frontier* (*SEF*), embodied the perspective of many in industry, as well as many among the science elite, and contrasted sharply with the position staked out by Kilgore in his two legislative efforts. Organized as a single report with the reports of the individual committees included as appendices, *Science—The Endless Frontier,* like Kilgore's legislation, called for the creation of a new scientific agency. In the same breath, however, the report expressed the concern of Bush and others that the establishment of a new agency to be called the National Research Foundation (NRF) not burden the scientific enterprise with too much bureaucracy. As the Bowman committee put it: the "primary purpose [of the proposed agency] is to provide encouragement, and where necessary, financial aid, without at the same time introducing centralized control of research" (Bush [1945] 1960: 116).[20] Of course, Kilgore's proposed agency aimed to centralize research policymaking, not research as such.

In the area of organizational governance, *SEF* advocated an organization "free from political influence" or, as the Bowman committee put it more strongly, "free from the influence of pressure groups, free from the necessity of producing immediate practical results, free from dictation by a central board" (Bush [1945] 1960: 51, 79). In practice, this meant the organization should be controlled by persons "having an understanding of the peculiarities of scientific research and education," that is, scientists (9). These persons would be appointed to the NRF board by the president, and it would be up to the board itself to select the foundation's director. The Bowman committee went so far as to recommend that the president select his choices for the board from a group nominated by the pinnacle of elite science in the United States, the National Academy of Sciences (115, 35). This form of organization, of course, contrasts sharply with Kilgore's second proposal, which called for a board composed of representatives of a broad range of social interests and a director appointed by the president. The differences between the governance mechanism proposed by Kilgore and that proposed by Bush ended up being central to the legislative debate over postwar research policy. Bush saw his proposed arrangement as fundamentally in keeping with the principles of American democratic government (33); however, his proposal raised long-standing issues of executive prerogative and the accountability of appointed officials to elected officials and thus to the voting public.

Unlike Kilgore's proposed agency, in line with the position taken by industry, Bush's National Research Foundation aimed to *support primarily basic research.* New products and processes, the report asserts, following a logic urged by those in industry, "are founded on new principles and conceptions which in turn result from basic scientific research." Basic research "creates the fund from which the practical applications of knowledge must be drawn" (Bush [1945] 1960: 6, 19). The report further justifies an emphasis on basic research on the grounds that advances in basic research "when put to practical use mean more jobs, higher wages, shorter hours, more abundant crops, more leisure for recreation, for study, for learning how to live without the deadening drudgery which has been the burden of the common man for ages past" (18).[21]

While Bush may have been quite sincere in his claim that basic research must form the basis for applied work and the benefits which will flow from applied work, this assertion has a hegemonic character. In Bush's formulation, scientists' interests lie in the opportunity to undertake unencumbered basic research; the public, on the other hand, clearly has an interest in receiving some material benefit—an improved standard of living—from their government's investment in research. By claiming that basic research forms the foundation for applied work, Bush tied the interests of scientists in doing basic research to the interests of the general public in the material benefits of science.

Although the Bush report contends that the best way for government to promote industry research is indirectly through supporting basic research, it does admit some more applied role for the proposed National Research Foundation. Indeed, although the foundation would not be a protoindustrial policy agency of the variety Kilgore advocated, both the main report and the Bowman committee report recommend the foundation "devise and promote the use of methods of improving the transition between research and its practical application in industry" (Bush [1945] 1960: 37, 75).[22]

Later proposals supported by Bush suggested only a limited policy-making role for a federal science agency. The Bush report, however, calls for an advisory body to *coordinate* the science policies of all government agencies. Of course, Kilgore consistently put a premium on policy coordination. Also like Kilgore's proposals, the Bush report calls for a single national policy for science ([1945] 1960: 7, 31).

A final matter of disagreement between the Kilgore proposals and the Bush plan was in the area of patent policy. In the report, Bush admitted what Kilgore had asserted, that there had been abuses in the patent system. But he continued to claim that the patent system was "basically sound" (Bush [1945] 1960: 21). In the report, however, he was more specific than he had been in his letter in *Science*. In the report, Bush challenged Kilgore's position that inventions based on government support should all be licensed on a nonexclusive basis, arguing to the contrary that "there should certainly *not* be any absolute requirement that all rights in such discoveries be assigned to the Government, but it should be left to the discretion of the Director and the interested Division whether in special cases the public interest requires such assignment" (38).

The Bush report highlighted several other issues, not raised in Kilgore's early legislation, that were to become important in the following legislative debate. First, the report argued for a continued civilian role in military research in the postwar period (Bush [1945] 1960: 6, 18, 34). Second, while the *SEF* committee on medical research recommended establishing a separate medical research foundation, in his portion of the report, Bush called on his NRF to support medical research (35, 47). Finally, the Bush report stressed the importance of government support of basic science in colleges, universities, and research institutes (20).[23]

According to one historian, "the Bush report was widely quoted in the press and quickly overshadowed the . . . [work] of Senator Kilgore" (McCune 1971: 80; see also B. L. Smith 1990: 42–43). This is almost certainly true, but it is important to understand why. To do this, we must look at the intersection between the structure of the polity and broad symbolic capital enjoyed by scientists—Bush in particular—as the war came to a close. Roosevelt's support for Bush was central; indeed, as I have noted, the president went so far as to ask a staff director of one congressional committee to postpone hearings on science policy until after Bush issued his report.[24]

The relationship between Bush and Roosevelt depended on a particular historical conjuncture and a particular state and party structure. The Roosevelt administration proposed some of the most radical organizational reforms in U.S. history. Kilgore's proposals were in line with administration initiatives, and Bush's ran counter to many New Deal principles. Despite their ideological compatibility, Roosevelt and Kilgore do not seem to have been particularly intimate.[25] Perhaps Bush played on President Roosevelt's progressivism—his belief in the corruption associated with big government. But that Bush had such a good relationship with Roosevelt was, more important, a product of the war situation in which the president depended inordinately on Bush, and, as Bush's agency succeeded, Roosevelt came increasingly to trust and respect Bush. The symbolic capital of science grew.

While the war situation may have facilitated the closeness of the Bush-

Roosevelt relationship, the structure of the American polity must also be considered. There was little discipline within the administration, and, indeed, there was substantial division over whether Bush's plan or Kilgore's should be supported. In this context, Bush's personal relationship with President Roosevelt is what mattered most in defining the administration's position.

In the absence of party discipline, Kilgore was not obliged to go along with the president's support of Bush. Perhaps this would have settled the issue quickly. However, even before *Science—The Endless Frontier* was issued in July 1945, Roosevelt died, and Bush lacked the same intimacy—indeed, the same credibility—with Truman (Bush 1970: 293, 303; Kevles 1987: 362). Bush urged Truman to submit the report to Congress with his endorsement, but, instead, Truman submitted the report to Congress for further study. This situation—a product of timing and extensive fragmentation and a lack of party and bureaucratic discipline—provided the foundation for a long and drawn-out struggle.

Conclusion

The intersection of a range of structural and conjunctural factors served to define the character of Kilgore's and Bush's positions in the debate over the postwar organization of research policymaking in the United States and the relative advantages and disadvantages of Kilgore and Bush. Kilgore's agenda was a product of his interpretation of the wartime domestic situation filtered through populist and New Deal discourses. His own biography was rooted in an antibusiness and small *d* democratic populism, and his program emphasized the needs of the individual inventor, the small businessperson, and the average consumer. His solution to the problems induced by business elites and bureaucrats was not to turn the problem over to elite scientists, but to promote democratic control and a New Deal program of coordination and planning.

Kilgore's institutional soapbox—his role as chair of the Senate Subcommittee on War Mobilization—provided him a place from which to speak. From there, he was able to request reports and hold hearings that would bolster his position. On the other hand, the New Deal discursive legacy on which he relied was not firmly rooted in broader American discourse. Its legitimacy was tenuous, and Kilgore was forced to confront an ideology of science, the credibility of which had grown tremendously with Allied success in the war. Then, too, Kilgore faced a fragmented state, a divided executive, and a Democratic Party not equipped to promote party discipline. As a result, despite Democratic control of the Congress and executive, rapid passage of Kilgore's early legislation was not possible.

Bush's plan was fundamentally shaped by a long-standing scientists' discourse: a view of science as appropriately autonomous from society and above and outside of political affairs. The war had bolstered the symbolic capital of science and this perspective in particular, and Bush was able to use this discourse as a means to delegitimize Kilgore's efforts to promote public control of scientific research resources. Moreover, in keeping with this viewpoint Bush attempted to show that what Kilgore viewed as the coordination of science policies and research efforts amounted to a kind of regimentation of science that was inimical to the norms of science themselves.

What is more, Bush and his colleagues had earlier used their credibility and their social capital to obtain a place in the state from which to direct the science and technology component of the U.S. war effort. Scientists' war work had further enhanced their symbolic capital and Bush's and his colleagues' social capital. Bush's intimate relationship with President Roosevelt provided the opportunity and institutional location from which elite scientists and their business allies could contribute substantially to defining the terms of debate and later influence the further course of debate.

With the initial terms of debate established, the structure of the American state and the character of political parties became central to the post-1945 course of postwar research policy legislation. To understand the post-1945 trajectory of U.S. research policy legislation, the ever narrower contours of a *central* postwar research policy agency and the related institutional fragmentation of postwar research policymaking, we must turn our attention to the organizational structure of the U.S. state and the related organization of political parties and social interests. This is the focus of the next chapter.

5

TOWARD PEACE ON THE POTOMAC

State Building and the Genesis of the National Science Foundation

---

> The real answer is that we must get a Foundation."
>
> Vannevar Bush, November 14, 1946

> I believe there should be a central agency—to be known as the National Science Foundation—which would have sufficient funds at its disposal to be able to assist other organizations, public and private, in carrying out research programs that are essential for national prosperity and welfare as well as national defense.
>
> Senator Harley Kilgore, February 5, 1945

> *Laissez-faire* has been abandoned as an economic principle; it should also be abandoned, at least as a matter of policy, in science.
>
> Waldemar Kaempffert, journalist, *New York Times,* 1943

> Everything possible should be done to direct public attention to the importance of keeping science and research free of government domination.
>
> R. J. Dearborn, chair, National Association of Manufacturers, Committee on Patents, April 11, 1945

With few exceptions,[1] there was broad agreement after World War II that some form of government mechanism was necessary to provide support for scientific research. There was little agreement, however, on the nature of this mechanism. The issues that divided Bush and Kilgore generally

Table 5.1 Basic Differences in Proposals by Kilgore and Bush

	Populist Proposal (Kilgore)	Scientist/Business Proposal (Bush)
Coordination/planning	Strong mandate	Vague coordination mandate
Control/administration	Business, labor, farmers, consumers	Scientists (and other experts)
Research supported	Basic and applied	Only basic
Patent policy	Nonexclusive licensing	Effectively no nonexclusive licensing

defined the broader terrain of debate over the establishment of a postwar science agency (table 5.1).

Five central issues divided the parties to the debate (England 1982: 5; U.S. House Task Force on Science Policy 1986a: 24). Probably the most important was how the organization should be controlled. Kilgore and his allies supported an organization under lay control and responsive to the president (Study Group, Washington Association of Scientists 1947: 385). Bush and his allies advocated an agency controlled by scientists (Kevles 1987: 346). Debate also centered on what type of research should be funded by this government body. Bush favored an organization that supported primarily basic research, while Kilgore believed that science of direct economic and social importance—applied science—should also be supported (Kevles 1977: 16, 22). In addition, Bush opposed including support for the social sciences under the proposed foundation's mandate. Important disputes also emerged over whether the government or the scientist should retain title to inventions developed with government support and over whether there should be an attempt to distribute resources evenly across the United States or whether funds should just go to the "best science" no matter where it was being done (Kevles 1987: 151).[2] Finally, the question of the extent of this government body's policy and coordinating and planning role became a matter of some dispute during the course of congressional debate.

Given the extent of division over the nature of a postwar research policy agency, it is perhaps not surprising that resolution was not immediately forthcoming. But the mere extent of disagreement is not sufficient explanation for the five years and four distinct attempts it took before legislation was finally signed into law establishing the National Science Foundation. As I argued in earlier chapters, an understanding of American research policymaking requires that we move beyond proximate causes to more general structural and historical causes.

In this chapter, I explain how and why certain actors were able to influence legislation, and I provide an organizational explanation for the lengthy delay in passing legislation. In brief, I argue that the permeable character of the Congress allowed actors to influence the substance of legislation on the basis of organizational position, social connections, and credibility, or some combination of all three. More fundamentally, I argue that the five-year delay is best explained by the highly permeable and divided character of the American state, the fragmentation of congressional authority, and the lack of discipline in the Democratic party. Easy entry to the state gave social interests the capacity to stop legislation at crucial points. At the same time, despite Democratic control of the legislature and executive throughout most of the period, lack of Democratic party discipline and the powerful positions held by congressional Democrats representing the southern United States slowed the legislation down. Finally, the delay—itself the product of a weak and fragmented polity—led to a highly fragmented configuration of organizations for the establishment of government research policy.

Context for Struggle: The Structure of the State and Civil Society

Some historians explain the delay in passage of NSF legislation in terms of "differences in personalities and individual rivalries" rather than differences over issues and party politics (England 1982: 108). Others blame Bush's persistence in proposing a foundation "so far above politics to be politically unrealizable" for the delay (Forman 1987: 183). More generally, historians explain the delay in terms of congressional votes and presidential vetoes. There is certainly truth in such explanations. But they beg more fundamental questions. Why did individual rivalries have such force? Why was Bush's plan unrealizable?

Understanding the viability of a particular legislative proposal, the relative importance of individual rivalry, as well as congressional votes and presidential vetoes, is enhanced by viewing the struggle over NSF legislation in a broad sociohistorical context. The power of individuals is particularly important within a context in which social interests are poorly organized and the state is highly permeable. By contrast, where negotiation between states and peak associations is institutionalized, individuals can have less autonomous impact on policy. In a state with a unified executive, centralized congressional power, and a disciplined Democratic party, Bush's plan would either have been quickly accepted and enacted or more likely dismissed in favor of Kilgore's program. Finally, while I would not deny the importance of committee votes and a presidential veto in explaining the delay in passage of NSF legislation, these are only *proximate* causes, and their significance is better understood by considering them in relationship to the structure of the American state and its relationship to civil society.

The structure of the U.S. state and its relationship to the organization of social interests in the United States has an "inherent historicity" (Skocpol 1985: 28). It is the organization of the state and civil society that constitutes the sociopolitical context within which to understand the debate over postwar research policy, and by attending to and historically situating this context, I attempt here to historicize what would otherwise be an ahistorical explanation.

I begin with the American state. The American state is perhaps the archetypal fragmented state. At the national level, the organization of policymaking has several dimensions. Within the Congress, committee jurisdictions overlap. Within the executive, agency decision-making authority is duplicative. Policy can be made through legislation, regulation, or executive order. Policies can be initiated by the Congress or the executive.[3]

Multiple points of entry, overlapping jurisdictions, and multiple veto points make the U.S. state highly permeable. In this context, social interests can gain access to policymakers relatively easily.[4] In addition, the line between state and society is blurred. What is more, the fragmented character of the U.S. state means that on any given policy issue there are multiple points of access. If an interest group does not get satisfaction from one congressional committee, it can turn to another or to an executive agency in an effort to influence policy. Finally, dispersion of authority and overlapping jurisdictions means that there are multiple points at which policy proposals can meet their demise. They may be tabled in congressional committees and prevented from coming up for congressional vote. They may be vetoed by the president or implemented by executive agencies in a way not in keeping with legislative intent.

Authority is dispersed through the federal system from the national state to city governments. As Skocpol argues, these factors and "the close symbiosis between the federal administration and Congressional committees all help ensure that state power in the twentieth-century is fragmented, dispersed, and everywhere permeated by organized social interests" (1985: 12). Shonfield describes this state of affairs as "riotous pluralism" (1965: 323).

The U.S. state lacks esprit des corps among civil servants and authoritative planning agencies (Shonfield 1965: 318–26; Skocpol 1985: 12). As Atkinson and Coleman note, a consequence of this "bureaucratic pluralism" is "incremental, short-term decision-making that is based on the lowest common denominator criteria and always vulnerable to the introduction of a partisan political calculus" (1988: 5; see also Shonfield 1965).

The U.S. state is a weak state. Reference to state strength in this context typically does not refer to national defense or security capacity,[5] nor, as I use the term here, does it refer to influence in foreign affairs. Rather, state strength points to a national government's capacity to develop and implement clear, comprehensive, and coordinated policies. With such capacity often comes a long-run planning orientation. In economic policy, weak states are typically restricted to macrolevel fiscal and monetary interventions, whereas strong states can influence the economy more selectively, focusing on industries or sectors. Fragmented and permeable states, like the United States, are typically weak.[6]

An important result of the divided authority between the executive and the Congress has been a long history of struggle between the two over issues of administrative expansion. Early attempts to create administrative agencies were promoted by the executive but opposed by the Congress, and when agencies were created, struggles between the president and the Congress over control ensued (Shonfield 1965: 320). In this context, even during periods of crisis like World War I and II, *"no centrally coordinated, executive-dominated national bureaucratic state" had much chance to develop* (Skocpol 1980: 175; emphasis added).

The highly fragmented character of the American state is the product of historical struggles (Skowronek 1982). In the early twentieth century, Progressive reformers attempted to weaken the control of patronage-oriented political parties over policymaking. They aimed to replace the power of patronage with expert controlled agencies at all levels of government. These reformers, however, feared significant expansion of the government's capacity to spend, and they did not favor significantly increasing the power of the national government. Moreover, as Weir and Skocpol conclude, "the successes of the Progressive[s] . . . were scattered and incomplete, and their partial success combined with the weakening of party competition in the early twentieth century United States to exacerbate tendencies toward dispersion of political authority within the American state structure as a whole" (1985: 135).

Not only is the U.S. national state weak, but so are American political parties. Historically, political parties in the United States have been locally oriented. Their main purpose has been "organizing elections and succession in office" (Weir 1988: 187; see also Shefter 1977 and Skocpol 1980: 195). This role has contributed to their patronage orientation.

American political parties generally have not played a role in policymaking (Lowi 1967: 255, 276). Given their local patronage and election orientation, this is not surprising. Weak at a national level and focusing on diverse local appeals, political parties in the United States are not in much of a position to develop national policies, and implementation of party programs is made nearly impossible, given the nonprogrammatic focus of parties and their related inability to exert discipline on party members to adhere to a national program. As Shefter notes, we lack a "responsible party system" in the United States in which a president elected by a majority can extend his/her sway over Congress (1978: 242).

Efforts have been made to nationalize American political parties and establish a modicum of discipline (Leutchenburg 1963: 266, 269), but without success. In the absence of a system of disciplined programmatic parties, "once in office, parties exhibit some discipline . . . , but members of Congress are remarkably free as individuals (or ad hoc coalitions) to pursue whatever legislation or whatever administrative measures they believe will appeal to local constituents or to organized groups of financial contributors and voters" (Skocpol 1980: 195). Ad hoc deals between opposing party factions are possible, and support for one policy is not necessarily tied to opposition on another. There are no institutional sanctions for alliance or coalition switching, and in this context, policy is not the product of coherent foresightedness, but of a series of ad hoc tradeoffs (Weir 1988: 187; Weir and Skocpol 1985: 145).

Of course, not all party systems are like the U.S. system. A particular historical dynamic produced the characteristics of the United States' party system. In most European countries bureaucratization in the state preceded electoral democracy. In this context, parties were not in the position to offer patronage in return for support of voters and party regulars because access to government jobs was controlled by bureaucratic elites (Shefter 1977). As a consequence, "parties were forced to rely upon programmatic appeals, based on ideology or promises about how state power might be used for policies advocated by, or potentially appealing to, organized groups of constituents" (Orloff 1988: 43–44).

In contrast to the European case, in the United States electoral democratization preceded bureaucratization of the state. As a consequence, parties could use patronage—offers of government jobs and resources—as enticement for support. There was no bureaucratic elite controlling access to the spoils of government. In this situation, as Orloff notes, "parties . . . tend[ed] to rely on patronage rather than on programmatic or ideological appeals to mobilize their constituents and reward activists" (1988: 44).

Given what I have said about the organization of the state and political parties in the United States, it should come as no surprise that social classes, fractions, and groups are generally poorly organized in the United States. Indeed, Skocpol has gone so far as to argue that the organization of the American state and American political parties encourages "a proliferation of competing narrowly specialized, and weakly disciplined interest groups" (1985: 23–24). Unions represent only a small proportion of American workers; the labor movement is not highly centralized and is not a major force in most national policymaking.

The American business community is heterogeneous and lacks a powerful peak association to speak on its behalf (Katzenstein 1978: 308). Organizational representation of business is, for the most part, decentralized, and national associations such as the National Association of Manufacturers are relatively weak (Katzenstein 1978: 311). Scientists, actors of political importance in the history of research policymaking, are organized into an array of professional associations in the United States. Additionally, scientists have organized such groups as the Federation of American Scientists with the explicit aim of influencing policy (Nichols 1974; A. K. Smith 1970).

Finally, given the highly permeable and fragmented character of the American state, the nonprogrammatic nature of American political parties, and the fragmented character of civil society in the United States, the role of *informal* contacts between state actors and elites and the role of individual elites in powerful locations within the state turn out to be central to understanding the process of developing a research policymaking institution in the United States in the postwar period.

Scientists, Business, and the Military

The dispute over postwar research policy is—if it is not portrayed simply as a conflict between two men (Bush and Kilgore)—often portrayed as a dispute between scientists and populists (B. L. R. Smith 1990: 40). In addition, while the alliance between science elites, big business, and the military is often noted, it is as frequently underplayed (Rossiter 1980: 551); indeed, the extent and importance of interlocking directorates and the strength of social ties in explaining the legislative outcome is generally overlooked altogether. Finally, while there is often recognition among historians of this period that the scientific community was not monolithic on the question of postwar research policy, the science vanguard headed by Bush is often seen to speak for the scientific community more generally (England 1982).

There was a relatively large group of scientists associated with nonelite universities often working in the biological and social sciences who supported Kilgore.[7] Often unorganized, these scientists generally opposed the concentration of research in a few universities and the concentration of the power to control research resources (Penick et al. 1972: 115–18). These scientists had allies in academic institutions that were not among the elite universities in the United States. Small and medium-sized institutions which had not benefited from the government's wartime largess—based largely on a logic that Bush saw as allocation based on merit, but which can easily be seen as based on a logic of who one knows—tended to support the more democratic distribution of resources implied in Kilgore's early proposals (Rowan 1985: 3).

As I discussed in chapter three, some of these nonelite scientists were members of the American Association for Scientific Workers (AASCW). Although dominated by nonelite scientists, the AASCW did include some prominent scientists (for example, Harold Urey) among its members. The AASCW was concerned with the democratization of science and the need for a high degree of coordination in research policy, and the organization was solidly behind Kilgore. However, the AASCW was relatively ineffective politically. The organization aided Kilgore in drafting his Science and Technology Mobilization legislation, but its members were not involved in OSRD-related projects and thus lacked important social capital and credibility. In addition, the organization was dominated by biological scientists who were not central to the war effort and consequently lacked the symbolic capital that enveloped scientists who were; the organization's focus was generally local, and members lacked the national connections necessary to influence policy (Hodes 1982; Kuznick 1987: 251).

Like the scientific community, American business was also divided on the issue of establishing a federal research policy agency. A national survey of two hundred manufacturers from across the country, undertaken in 1945 for Kilgore's subcommittee, found widespread industry support for government funding of research through a new central agency. While there was some variation between small and large businesses, a majority of both favored government support for basic research. Seventy-seven percent of those interviewed responded affirmatively to the question "Is federally supported scientific research needed by industry [for long-range use]" (U.S. Senate 1945/6: 385). There was less agreement on government support for industrial research. Seventy-two percent of small concerns favored government support for industrial research, as compared to 51% of large concerns. Neither large nor small businesses surveyed believed existing agencies should administer government sponsored research. The majority favored a new central science policy agency and believed research should be undertaken in existing government laboratories and by universities and nonprofit organizations on a contract or grant basis (U.S. Senate 1945/6: 386–93).

Although the results of the Kilgore survey do not indicate a perfect correlation between the position taken by industrial sectors in the debate

Table 5.2 Principal Industrial Contractors with the Office of Scientific Research and Development, 1941–45
(ranked by total dollar value of funds contracted)

1. Western Electric Company
2. Research Construction Company
3. General Electric Company
4. Radio Corporation of America
5. E. I. du Pont de Nemours and Company
6. Westinghouse Electric Manufacturing Corporation
7. Remington Rand, Inc.
8. Eastman Kodak Company
9. Monsanto Chemical Company
10. Zenith Radio Corporation
11. Standard Oil Development Corporation
12. Hygrade Sylvania (Sylvania Electric Company
13. Erwood Sound Company
14. Douglas Aircraft Company
15. M. W. Kellogg Company
16. Budd Wheel Company
17. Gulf Research and Development Company
18. Delta Star Electric Company
19. Emerson Radio and Phonograph Corporation
20. Ford, Bacon and Davis, Inc.
21. Globe-Union, Inc.
22. Federal Telephone and Radio Corporation
23. Bowen and Company
24. National Carbon Company
25. Galvin Manufacturing Corporation

Source: Baxter 1946: 456, 457.

and the structural characteristics of firms or sectors, contemporary accounts and historiographic material suggest that larger firms were most actively opposed to Kilgore and supportive of Bush. This makes sense since it was larger firms that tended to have their own research capacity. Many firms with internal research capacity opposed government support of industrial research because they feared competition with the government or perhaps competition from other firms that would rely on government-supported research (Redmond 1968: 117; Rowan 1985: 35). Conversely, government support of basic research was often favored by business because industrial support of basic research was not considered profitable (Davis and Kevles 1974: 219; Rowan 1985: 84).[8]

Business generally favored private appropriation of the results of government-sponsored industry research. A few select large firms obtained government contracts during the war (table 5.2) (Baxter 1946: 456–57), and, indeed, an important fear of business was that if Kilgore and his supporters had their way business contractors would not be entitled to exclusive rights to the products of the research they carried out with government support (Rowan 1985: 36). This would weaken an important advantage these firms had over their smaller competitors, and there was certainly widespread business opposition to Kilgore's efforts to prohibit exclusive licenses resulting from industry research supported with government funds.

In the legislative battle, the most visible organized business involvement came from the National Association of Manufacturers (NAM). In 1946, NAM claimed to have some 15,000 member firms who "produce[d] eighty-five percent of the manufactured goods in America."[9] By the mid-40s, the National Association of Manufacturers represented largely small concerns (Weinstein 1968: 92).

NAM's consistent opposition to Kilgore's proposals and support for Bush's efforts is a bit ironic, since Kilgore explicitly aimed to aid small businesses through his proposals to provide government support for industrial research and to require nonexclusive licensing arrangements. One could argue that the organization's opposition was primarily ideological. Certainly the evidence suggests NAM had a strong free market ideology and generally opposed any government involvement in the economy as a dangerous step toward socialism. This is no doubt part of the explanation, but, in addition, it is important that during the period under discussion the organization's leadership included significant representation from some of America's larger companies. Thus, in 1946 among the companies with representation on NAM's board of directors was American Cyanamid, duPont, Eastman Kodak, and Standard Oil.[10]

During the same year, the NAM Committee on Patents and Research—the group most responsible for formulating the organization's position on science legislation—included, among others, representatives from duPont, Westinghouse, and Radio Corporation of America (table 5.3).[11] These were the kinds of companies that had substantial in-house research capacity, and several of these companies had government contracts during the war. Thus, it makes sense that under their leadership NAM would be in the forefront of opposing Kilgore's proposals for government support of applied research and his early position on patent law, even if this opposition went against the interests of a significant portion of the organization's membership.

NAM took an active interest in the science legislation throughout the 1940s. They issued press releases (most often supporting Bush's ap-

Table 5.3 NAM Leadership among the Top Twenty-Five OSRD Contractors, 1940s

Firm	Rank by Total Dollar Value of Contracts Received
Eastman Kodak Company	8
E. I. du Pont de Nemours and Company	5
General Electric Company	3
Radio Corporation of America	4
Standard Oil Development Company	11
Westinghouse Electric Manufacturing Corporation	6

Sources: Baxter 1946 and Papers of NAM, Hagley Museum and Library, Accession 1411.
Note: These six firms together received roughly one-third of the total dollar value of contracts granted by OSRD to its twenty-five principal industrial contractors (see table 5.2) and were represented on the NAM board or on the NAM Committee on Patents and Research during 1942–50. Standard Oil of New Jersey was also represented on the NAM board and Standard Oil of Indiana was also represented on the NAM Committee on Patents and Research.

proach), the issue was extensively discussed by the organization's Committee on Patents and Research as well as in board meetings, and, of course, they lobbied members of Congress. The organization was generally supportive of the commitment of government resources to basic research, but NAM feared "political" control of a new government science agency and consistently opposed any provisions the organization believed would interfere with monopoly rights granted through patents. As R. J. Dearborn, chair of the NAM Committee on Patents and Research and President of Texaco Development Corporation, stressed in 1945, "everything possible should be done to direct public attention to the importance of keeping science and research free of government domination."[12]

NAM sometimes complained about its ineffectiveness in directly influencing Congress on this issue. Minutes from a June 1946 meeting of the NAM Committee on Patents and Research report that "our views have repeatedly been ignored" by Congress on the issue of the role of the proposed National Science Foundation in determining property rights.[13] There is no doubt that the organization's formal status as a business "peak association" might lead some representatives and senators to pay attention to the organization's position, but NAM's direct organizational influence on the debate was limited.[14]

The United States lacks corporatist arrangements in which relations between the state and peak business associations are institutionalized and business is given a formal voice in policymaking. In the United States, an organization's political influence may be facilitated by the material resources it can amass, but its organizational influence is not institutionalized. Instead, given the permeable character of the American state, informal relations with state managers and opportunities for representatives of business interests to serve as state managers are more viable means through which business can influence policy. Thus, in the American state social capital is often a more important resource than money or organization in influencing policy.

On the one hand, NAM appears largely to have lacked the social capital necessary to influence NSF legislation. On the other hand, the organization's complaints may be overstated. NAM's direct influence was certainly limited, but the organization's indirect influence and that of business more generally was certainly significant. Well before science legislation was debated in the Congress, Karl Compton, a close Bush aide and president of MIT, chaired the NAM's Research Advisory Committee. With the OSRD staff, Compton shaped proposals for science legislation, and certainly his early experience with NAM affected his perspective.[15] Of course, Bush's role in a variety of companies surely influenced his own views on such issues as possible patent provisions of proposed science legislation. Thus, the influence of business on science legislation was, to an important extent, *indirect,* through shaping of the perspective of members of the science vanguard, who themselves had the social connections and credibility necessary to shape policy.

A range of other informal and indirect means of business influence may have existed, but a central mechanism was through the Directors of Industrial Research. Formed in 1923, the Directors of Industrial Research (DIR) group had its origins in a meeting of the business-oriented offshoot of the National Academy of Sciences, the National Research Council. DIR was a small organization. Including emeritus members, the group had fewer than sixty participants. The ostensible purpose of the organization was to meet on a monthly basis to discuss issues of common interest such as industrial laboratory operation. Meetings were held at prominent New York City clubs including the Engineers' Club, the City Club, the University Club, and the Century Club.[16] DIR records stress the informality of the organization. As Merck research director Randolph Major remarked in a letter, "the group has always been considered extremely informal and no publicity has ever been released, as far as I know, on the organization."[17] Indeed, the group required no membership fee, except the cost of the monthly lunch.[18]

The records of the DIR indicate that the group took a consistent interest in the science legislation debate. As the organization's name implies, firms of organization members had internal research capacity. Thus, it makes sense that the Director of Industrial Research's position would generally follow the logic adhered to by large firms. They opposed government support for applied research, calling instead for a federal science foundation to focus exclusively on basic research. Predictably, organization members also opposed Kilgore's efforts to require nonexclusive licensing of inventions resulting from government supported research.

That the DIR appears to have had a greater influence on NSF legislation than the considerably more visible and highly organized NAM points again to the importance of informal connections and social capital in shaping policy within the context of a highly permeable state. The high point of DIR influence came in shaping a legislative proposal of Senator H. Alexander Smith. Smith, a New Jersey Republican, was linked to the group through his association with the New Jersey–based Merck company. Randolph Major, Merck's research director, played a leadership role in the Directors of Industrial Research.

In addition to this direct role, the DIR undoubtedly served various roles in the indirect shaping of policy. The DIR was connected to both Frank Jewett and Vannevar Bush. Jewett, the president of Bell Laboratories and the National Academy of Sciences, was a member of the Research Directors and kept members abreast of important developments in the legislative debates over the establishment of a National Science Foundation.[19] Bush was a member of the Merck board.

While I show specific and direct cases of business influence in this legislative struggle, it is important to recognize that the close and overlapping contacts between business and science elites like Bush and Jewett undoubtedly played a role in shaping the perspective of the Bush camp. It

Table 5.4 Overlapping Membership: National Association of Manufacturers and Directors of Industrial Research, 1942–50

Firms Represented in DIR	NAM Affiliation
Air Reduction Company	Member
Aluminum Company of America	Board member
American Brass Company	
American Can Company	Board member
American Cyanamid Company	Board member, Patent committee, Research committee[a]
American Viscose Corporation	Member
Armstrong Cork Company	Board member
Arthur D. Little	Research committee
B. F. Goodrich Company	
Bakelite Corporation	
Battelle Memorial Institute	Research committee
Bell Telephone Laboratories	
Carbide and Carbon Chemical Company	
Colgate-Palmolive-Peet Company	Board member
Corning Glass Works	Member
Dorr Company	
E. I. du Pont de Nemour & Company	Board member, Patent committee, Research committee
Eastman Kodak Company	Board member
Electro Metallurgical Company	
General Aniline and Film Corporation	
General Electric Company	Board member, Patent committee
Goodyear Tire and Rubber Company	Board member
Gulf Research and Development Company	Member[b]
Hercules Powder Company	Member
International Nickel Company	Member
Johns-Manville Corporation	Member
Johnson and Johnson	
Mellon Institute of Industrial Research	Research committee
Merck and Company	Patent committee
National Carbon Company	
National Lead Company	Member
New England Industrial Research Foundation	
New Jersey Zinc Company	
Owens-Corning Fiberglass Corporation	Member

Table 5.4 *Continued*

Firms Represented in DIR	NAM Affiliation
Park, Davis, and Co.	
Radio Corporation of America	Patent committee, Research committee
Schenley (Distillers) Corporation	
Standard Brands	
Standard Oil Development Company	Board member,[c] Patent committee[d]
U.C.C. Laboratories	
Union Carbide and Carbon Research Laboratories	Member[e]
United States Rubber Company	Board member
U.S. Industrial Chemicals	
U.S. Steel Corporation	Member
Westinghouse Electric and Manufacturing Co.	Patent committee

Source: Papers of the Directors of Industrial Research and Papers of the National Association of Manufacturers, Hagley Museum and Library.

[a]Full titles: Committee on Patents and Research; Research Advisory Committee.

[b]Gulf Oil Corporation.

[c]Standard Oil of New Jersey.

[d]Standard Oil of Indiana.

[e]Union Carbide and Carbon Corporation.

was then left to Bush and his colleagues, who were ensconced in the state, and for whom the war[20] had provided a network of social connections with legislators and a high degree of credibility in the eyes of the public and of elected officials, to actually promote the legislation.

If DIR influenced policy through shaping the views of the science vanguard, the perspective of the NAM leadership gained serious consideration through its members' roles in DIR and DIR's subsequent influence with Senator Smith. Not surprisingly, there was a great deal of overlap between the National Association of Manufacturers and the DIR (table 5.4). At least twenty-five of the forty-six companies represented in the DIR between 1942 and 1950 were members of NAM for some or all of the period. At some point during the period, representatives from seven of these companies sat on NAM's Patents and Research Committee, six served on the organization's Research Advisory Committee, and eleven sat on NAM's board of directors.[21]

While the group never took a direct stand, and there is no evidence that group members explicitly represented the DIR's interests to members of Congress, as I will show, there can be no doubt that business views were taken into account in drafting various iterations of science legislation. What is important to note here is that the likely mode of information transmission was informal. The overlapping and interlocking nature of memberships suggest that the formation of policy positions among powerful elites was likely to be indirect and the product of repeated interaction between involved actors.

A third group of actors with an active interest in postwar science legislation was the military. Both the U.S. Army and the U.S. Navy desired to maintain control of military-related research and opposed efforts to promote civilian control of such research, as proposed in early efforts by Bush (Reingold 1987: 338; Rowan 1985: 50). In addition, both branches of the service opposed Kilgore's early patent proposals. The army and navy feared that such provisions "would raise the cost of, if not make impossible, industrial contracts for military research and development" (Kevles 1987: 345). And the military saw such research as central to its mission.[22]

Both Bush and Kilgore sought the support of the military in their efforts. In their respective attempts to forge allies, Bush had a distinct advantage: he had worked closely with the military during the war. Although the military opposed Bush's early promotion of civilian control over military research, they had a long-standing relationship with him, and they trusted him. With Kilgore, on the other hand, the military had had a less than happy relationship. As early as 1941, he cosponsored legislation to create a Department of Defense. Kilgore's proposed department would have consolidated the war and navy departments, and neither department liked what each saw as an incursion into its sphere of responsibility.[23] In addition, the interests of the science elite and the military were compatible. The war had made the military sympathetic to scientists' "ancient ideology of independence" (Greenberg 1967: 126). The autonomy granted scientists during the war had paid off handsomely for the military. On the other side of the equation, the military provided a pipeline to government financial support for scientific research (126–29).

Positions in the State: Bush and Kilgore

In the previous two chapters, I assessed the structural advantages Bush and his colleagues at OSRD had as a result of the organization's role in the war effort—the credibility successes brought and the social connections Bush solidified with legislators in his role as liaison between OSRD and Congress. These factors were central in Bush's role in the NSF debates and, indeed, in his ability to serve as a conduit for business and military, as well as scientist, interests.

By the time National Science Foundation legislation was introduced, the OSRD was nearly ready to wind down. From the end of 1946 to the middle of 1947, OSRD staff had been reduced from 177 persons to only 74 (Stewart 1948: 332), and Bush bemoaned the loss of OSRD as a resource in shaping legislation.[24] Yet the contacts Bush made during the war outlived the dwindling resources of the OSRD, and, although his views represented only those of a small science elite, Bush had become the accepted spokesman for the scientific community (Hodes 1982: 24–25). He worked closely first with Senator Warren Magnuson on science legislation and later with Senator H. Alexander Smith. In the case of Magnuson, in particular, Bush actively shaped the senator's role. The two had a social relationship separate from their professional interactions. According to Robert McCune, an historian who interviewed Magnuson, Magnuson and Bush "lived only one and a half blocks apart in northwest Washington and, as a result, often traveled and dined together in the Capitol area." Bush played on this relationship in eliciting the senator's early support and in promoting Magnuson's early leadership on the issue (McCune 1971: 81). Not only did Bush work with Senators Magnuson and Smith, but Bush's chief legislative aide at OSRD, John Teeter, worked closely with both Senators in drafting legislation and plotting strategy.[25]

Throughout the period, Bush was able to maintain close contact with the military in his role as chairman of the Joint Research and Development Board, a government board established by the secretaries of the army and navy to coordinate military R&D activities. Importantly, although Bush was not particularly hurt by the reduction in staff at OSRD, his position may have been weakened by the replacement of Roosevelt by Truman upon the former's death. Bush had the solid support of President Roosevelt; President Truman, on the other hand, had a long-standing relationship with Kilgore (see Maddox 1981). Indeed, early on he made Kilgore's program the position of the administration (Maddox 1979: 33; Kevles 1987: 356). Nevertheless, Truman did not enforce executive branch discipline, and he permitted Bush to speak out in opposition to the administration position on science legislation. In fact, division within the executive may go a long way to explaining the drawn-out character of the debates.[26]

If Bush's structural location made him a power to be reckoned with, Kilgore was not totally powerless. Indeed, early on in the Truman administration, administration officials, including budget chief Harold Smith, "formed a coalition against Bush with considerable influence in the White House" (Kevles 1987: 356). In general, Budget Bureau officials favored the planning approach advocated by Kilgore in his early proposals. Administration officials favored an organizational structure for the NSF that would make the organization's director and board responsible to the president. Consequently, there was strenuous opposition in the administration, and by Truman in particular, to Bush's proposals to limit executive control of the proposed science agency (Kevles 1977: 25). Budget director Smith was also supportive of the social sciences and worked with representatives of the Social Science Research Council.[27]

Outside the administration, Kilgore was not without resources. As *Fortune* magazine noted at the time, the "powerful" Kilgore subcommittee, was "one of the half-dozen largest in Congress, with a big staff and four years' steady draft of funds" (1946: 212). At the same time, Kilgore faced opposition not only from Republicans, but from within his own party. Kilgore belonged to the progressive New Deal wing of the Democratic party, and as several commentators have noted, the party was divided between this faction and conservative southern Democrats. Southern legislators feared potential federal control that might accompany government expenditures. As Weir and Skocpol have noted:

> Many of the New Deal programs introduced after 1935 . . . entailed the intrusion of the federal government into jealously guarded local terrain. . . . Many southern representatives in Congress had strong ties to landlords who had dominated the region's political and economic life for over half a century. Proposals to expand the purview of the central government or to trans-

fer local functions to Washington were strongly resisted by these people. (1985: 145)

The dispersion of authority and the seniority system permitted southern Democrats to control key committee positions in Congress (*Congressional Quarterly* 1982: 60; Domhoff 1990: 235), and working with Republicans, southern Democrats posed a real barrier to rapid passage of Kilgore's initiative. Although Kilgore's legislative proposals did not pose an obvious threat to local control, they may have raised general conservative fears of expanding central government and were viewed as extravagant by fiscal conservatives from both parties (John Conner interview, 9/26/90). With a distinct agenda in a divided party and not restrained by the kind of sanctions that a disciplined programmatic party might impose, southern Democrats were free to oppose Kilgore and in a position to do so effectively.

Double-Cross and Deception: Early Efforts to Establish an NSF

In the remainder of this chapter, I explore in detail the four legislative attempts to establish a National Science Foundation. I consider two major issues. First, for each legislative initiative I explore how the substance of proposed legislation was shaped through informal contacts within the context of a highly permeable Congress. Second, I seek to explain the lengthy delay in passage of NSF legislation. I suggest the delay can be explained in terms of division between the legislative and executive branches, divisions within the executive, dispersed authority within the Congress, and the lack of discipline in the Democratic party.

To understand the delay in passing NSF legislation, the four legislative efforts can be viewed as a comparison of similar events across time. The major independent variables that explain legislative failure in the first three efforts are structurally equivalent: the state is fragmented (and permeable), and the Democratic party is not able to impose discipline. In the final case, major structural changes do not unify the state or the Democrats, but a short-term rule change limits the effects of state and party division, permitting congressional passage of the NSF Act of 1950.

Kilgore believed that the success of his legislative efforts depended on support from scientists, and prior to the release of Bush's report, *Science—The Endless Frontier,* members of Bush's and Kilgore's staffs met to discuss draft legislation for a national science agency (Rowan 1985: 57). Believing that he and Bush were cooperating, Kilgore agreed to hold back legislation until after publication of the Bush report (Hodes 1982: 127; Maddox 1979: 32), but the extensive philosophical differences between Bush and Kilgore led Bush to seek alternative sponsors for OSRD-drafted science legislation (Hodes 1982: 127; Rowan 1985: 57–58). In the Senate, Warren Magnuson—a liberal Democrat liked by the president—sponsored Bush's legislative proposal, and in the House Wilbur Mills sponsored the OSRD-drafted measure (Rowan 1985: 57). Without Kilgore's knowledge, the Bush bills, which aimed to implement the proposals put forward in *Science—The Endless Frontier,* were introduced on July 19, 1945, the same day as the report's release (Maddox 1981: 166–67). Kilgore believed he had been misled. A Kilgore ally commented that the senator felt "double crossed" and was "mad as everything" (quoted in Maddox 1979: 33). He introduced his own National Science Foundation legislation on July 23rd.

The Magnuson bill, S. 1285, called for establishment of a National Research Foundation (NRF) run by a board appointed by the president without regard "to political affiliation." The board itself was to be responsible for choosing its own chair and vice chair, as well as for selecting the foundation's director. Among the six divisions to be included under the foundation were medical, physical sciences, and national defense divisions, each administered by a committee selected by the board based on recommendations from the National Academy of Sciences. The bill, finally, called on the NRF to promote "a national policy for scientific research and scientific education." S. 1285 said nothing about the nature of the policy, but did require the foundation to "correlate" its programs with those undertaken by other organizations (U.S. Senate 1945b: 7–8). Later, an amendment suggested by Bush explicitly denied the foundation *supervisory responsibilities or a regulatory role* over other agencies involved in research policy issues (Jones 1975: 294) (table 5.5).

This piece of legislation staked out the battleground in the debate over control of the National Science (or Research) Foundation. Its emphasis on absence of political affiliation is in stark contrast to Kilgore's earlier concern with a board representing a wide spectrum of social interests. At the same time, the Magnuson bill took control away from the executive and put it in the hands of an unelected board. Although nothing in the legislation mentions a board composed of scientists, the requirement to consult with the NAS on committee appointments pointed in the direction of *scientist control* of the foundation. Finally, the Magnuson bill did not include discussion of other contentious issues, including support for the social sciences, patent policy, and geographical distribution of resources.

Although Kilgore's first science foundation bill, S. 1297, differed from Bush's bill on each of the major points, it did indicate Kilgore's desire to obtain support from the scientific community (Chalkley 1951: 12). Kilgore's 1297 was more moderate than his earlier mobilization bills, but retained several of the same themes. The director was still to be appointed by the president. Instead of several boards with public members representing specific social interests (e.g., labor and agriculture), the National Science Board which would advise the director was to have, in addition to the heads of several executive departments, eight "*public members.*" The bill provided no specification of the interests they should represent, and indeed the Science and Technology Mobilization bill was the last bill specifying the interests to be represented.

Perhaps in an effort to ease scientists' fears that a government science foundation would mean government meddling in science (Kevles 1977: 15), unlike earlier bills, 1297 did not permit the science agency to operate its own labs. On the other hand, the bill retained an expectation that the agency would support applied research, including pilot production facilities. The bill also gave the government patents resulting from government-funded research and authorized the agency to grant nonexclusive licenses on them (except where granting such a license could in some way provide the basis for monopoly).

The bill provided specific instructions on the percentage of support to military and medical research and proposed to establish distinct divisions within the agency for each. In keeping with Kilgore's earlier concern to establish *a central dominant federal science agency,* the bill also provided the agency with *an evaluation and recommendation role* (a research policymaking role) concerning all federally financed research, as well as a role *coordinating* government research. After surveying government research activities, the foundation was expected to recommend necessary changes.[28] The bill did not specifically mention the social sciences. Nor did it specify

Table 5.5 First Legislative Round: Kilgore versus Magnuson Bills

	Kilgore Bill (S.1297)	Magnuson Bill (S.1285)
Presidential appointment	Board and director	Board only; board selects director
Coordination/ planning	Vague mandate	Vague mandate
Control/ administration	Public members and civil servants	Probable domination by scientists (and other experts)
Research supported	Basic and applied	Primarily basic
Patent policy	Nonexclusive licensing	No discussion
Social sciences included	No discussion	No discussion
Geographical distribution of funds	No discussion	No discussion

requirements for geographic distribution of financial support (U.S. Senate 1945b) (table 5.5).

Business support for science legislation, as reflected by the views of the National Association of Manufacturers (NAM), was lukewarm at this stage. The association's board concluded:

> Because of the present conditions, some form of financial assistance for basic research was inevitable. Consequently rather than have the NAM appear in opposition to a proposal which in one form or another was certain of enactment, it was deemed advisable to ascertain what the most satisfactory method of such assistance by the government would be and to support a proposal along these lines. As a result, we are generally in favor of the Magnuson Bill and opposed to the Kilgore bill.[29]

NAM's patent advisor, George Folk, especially stressed opposition to the patent provisions in the Kilgore bill.[30]

Although Kilgore had attempted to garner support from the military through proposing creation of an independent military division in his agency, the military opposed the Kilgore bill (Rowan 1985: 62). They opposed the patent provisions in Kilgore's legislation, and in addition, they opposed the Senator's attempt to *coordinate* all government research through a *single agency* (82). This created something of a rift in the administration: Truman and others in the executive supported the Kilgore bill and opposed the Bush program since it challenged executive prerogative by giving the board the right to select the director and ultimately determine national science policy and the expenditure of public funds (Schaffter 1969: 11).

Truman's New Deal–oriented Budget Bureau strongly supported an administrative structure along the lines specified in the Kilgore bill. Bureau officials believed Kilgore's bill would make the foundation accountable to the voting public through the president, since the president was empowered to appoint the agency's director and board. In line with the planning tradition of the New Deal, the Bureau also supported an important coordinating role for the NSF in opposition to the military's position. Such a coordinating position would undoubtedly have enhanced the power of the bureau at the expense of other executive agencies, including the War and Navy Departments. Finally, although they generally supported the patent provisions in the Kilgore bill, the Budget Bureau was concerned about a provision giving the Justice Department responsibility for defending licensees of government-held patents from legal challenges and requiring the Justice Department to determine if patented discoveries would result in monopoly control (Rowan 1985: 68).

Organizationally, the scientific community was divided into two groups on the legislation. Those closely associated with the OSRD generally supported the Bush-inspired Magnuson bill (Rowan 1985: 88). Behind them stood a larger group headed by Isaiah Bowman, the chair of one of Bush's *Science—The Endless Frontier* committees. This hastily organized group, the Committee Supporting the Bush Report, was founded several months after the legislation was originally introduced, specifically to promote creation of the kind of agency outlined in the Bush report. The committee claimed to speak for the majority of U.S. scientists (England 1982: 37). In support of its position, the committee collected some 5,000 signatures from working scientists (*Fortune* 1946: 242).

Politically liberal scientists and many outside the OSRD loop favored the Kilgore measure, believing that the senator's proposed administrative structure would protect it from military domination and special interests.

Two prominent scientists who took this view were Harlow Shapley of Harvard and Harold Urey, who had done atomic bomb–related research. Shapley and Urey also feared the spread of secrecy in science and thought Kilgore's measure was more likely to avoid this perceived danger (Rowan 1985: 94).

In December 1945, Shapley and Urey formed the Committee for a National Science Foundation. The group's concern was less to support a particular measure than to promote compromise in the interest of resolving the controversy over a government science foundation. The two scientists obtained the support of two hundred prominent scientists, including five Nobel prize winners and such prominent figures as atomic scientists Einstein, Fermi, and Oppenheimer (England 1982: 40; *Fortune* 1946: 242).

Diversity of opinion was revealed in joint hearings held by Kilgore and Magnuson (Chalkley 1951: 19–20). To obtain the support of scientists and others who supported the Magnuson bill, Kilgore staffer Herbert Schimmel worked with Budget Bureau officials to craft a compromise measure (England 1982: 28; Maddox 1981: 173; Parsons 1946: 655; *Steel* 1946: 58).

The need for compromise legislation reflects the organization of the state and political parties. An insulated state, a disciplined Democratic party in the majority, and support of the president would likely have guaranteed passage of Kilgore's first bill. But a permeable and divided state and a divided Democratic party with no capacity to impose discipline meant passage of legislation would require giving a little bit to everyone.

The compromise bill, S. 1720, was introduced a few days after Shapley and Urey formed their committee in favor of a compromise. *Steel* magazine declared the new legislation "more conservative than the measure . . . [Kilgore] originally drafted" (1946: 58). In an attempt to satisfy both the administration and elite scientists, the foundation called for in the legislation was still to be controlled by a presidentially appointed director, but the director was to work closely with the board.

Kilgore compromised also on the bill's patent provision. Attempting to address his concern for guaranteeing the broadest possible access to inventions and the concern of Bush, as well as business leaders, that the provisions not hinder economic development, 1720 required patents that resulted from government-financed research to be freely available to the public, but gave foundation administration discretion in the assignment of patent rights resulting from contracted research. In addition, the proposed foundation would give considerable discretion over military research to the military and would focus primarily on basic research, not applied (Rowan 1985: 103). Finally, addressing the concerns of have-not universities, the bill made a commitment to geographical distribution of research resources: one-quarter of the foundation's funds were to be distributed according to a formula that divided two-fifths of it among all of the states equally and the remaining three-fifths of it according to state population. In both cases, money was only to go to tax-supported institutions (England 1982: 39; *Steel* 1946: 58, 60).

When the congressional session ended in December, S. 1720 died. A similar bill was introduced as S. 1850 early in 1946 with Senator Magnuson as a cosponsor. Bush, Magnuson, and Kilgore cooperated in crafting the compromise measure (Chalkley 1951: 16; England 1982: 41). Again, Kilgore made compromises he saw as necessary to win the bill's passage, especially to vocal scientist opponents (Maddox 1981: 232).[31] This new bill still gave the president the right to appoint the director as Truman demanded, but it required the president to consult with the board before making the appointment, a concession to Bush and his colleagues (Maddox 1979: 36). The bill contained a patent provision and geographical distribution of resources provision similar to the earlier Kilgore compromise measure (McCune 1971: 103–4). The bill also contained a provision for the creation of a social science division (Maddox 1979: 36).[32]

S. 1850 was reported out of the Senate Committee on Military Affairs—the committee of which Kilgore was a subcommittee chair—in April 1946. In reporting out the bill, Republicans on the committee issued a minority report signed by all committee Republicans except Senator H. Alexander Smith from New Jersey (McCune 1971: 103). When the bill reached the floor, Smith—possibly with the illicit support of Vannevar Bush—introduced a series of amendments to 1850, which amounted to the reintroduction of the original Bush-inspired Magnuson bill (Rowan 1985: 119). With the aid of other Republicans and southern Democrats, Smith attempted to alter the administrative structure of the proposed foundation, but failed on a tie vote. He also failed to alter the patent provisions of the bill, but was successful in eliminating provisions for a social science division in the proposed foundation (K. Jones 1975: 321; Rowan 1985: 119). The bill passed the Senate on July 3rd.

Bush opposed the administrative provisions of the compromise bill, and NAS president Frank Jewett believed the legislation would lead to socialization of "a large and independent section of our economy" and would mean domination of the "life of the nation by a small group of federal officers and bureaucrats."[33] But outside the elite inner circle of the scientific community, there was widespread support for the legislation among scientists. Indeed, a survey by the American Association for the Advancement of Science (AAAS) found overwhelming support among scientists for the proposed foundation (England 1982: 30). Moreover, a wide cross-section of scientific and academic organizations threw their support behind the legislation, among them earlier supporters of two formerly opposed scientific groups—the Committee Supporting the Bush Report and the Committee for a National Science Foundation (Rowan 1985: 110).

Outside the scientific community there was also widespread support for the legislation. Within the administration even the military—although concerned about the bill's patent provisions—gave their grudging support to the measure. The industrial press also expressed concerns but supported the bill (Rowan 1985: 110–11). The National Association of Manufacturers, however, while recognizing the compromises made in the bill, was not satisfied; they argued that "research would be seriously hindered through unwarranted Government competition" if the bill passed, and the organization's Committee on Patents and Research went on record in March 1946 as opposed to the compromise bill.[34]

American state structure lacks any mechanism to enforce legislative compromise. In the period between the day S. 1850 was reported out of the Senate Committee on Military Affairs and the time it was passed by the Senate, it became clear to Bush and others that Republicans and southern Democrats might be able to block 1850 in the House (Rowan 1985: 115). Both Kilgore and Magnuson had assumed Bush supported 1850, but Kilgore found himself again double-crossed when Representative Wilbur Mills, at the urging of Bush and with the assistance of Bush aide John Teeter, introduced House bill 6448 in mid-May. The bill was a barely modified version of the original Bush-inspired Magnuson-Mills bill. It reintroduced the administrative provisions favored by Bush, which left the foundation controlled by scientists and largely independent of the executive branch. In addition, the bill eliminated provisions for support of the social sciences and eliminated definite patent provisions. Finally, the bill set aside funds specifically for national defense and stressed the principal of support for "best science" over geographical distribution (115).

Introduction of the Mills bill destroyed the fragile coalition that had originally backed the Kilgore-Magnuson compromise. Although Truman and the Budget Bureau reaffirmed their support for 1850, high-ranking military officials threw their support behind the Mills bill (Rowan 1985: 116). The Committee Supporting the Bush Report also expressed support for the Mills bill, and Bush aides and supporters lobbied behind the scenes in an effort to assure its passage.[35]

A slightly modified version of the new Mills bill was reported out of the Subcommittee on Public Health of the Committee on Interstate and Foreign Commerce to the full Committee. On July 19, 1946, "the Committee decided it did not have the information necessary for action and the whole subject was tabled, thus blocking any further action before a new session of Congress" (Parsons 1946: 656).

There is some dispute over why NSF legislation died in the 79th session of Congress. One Bush aide blamed Frank Jewett for "putting S. 1850 to sleep."[36] Another contemporary suggests that legislators found early division between the Committee Supporting the Bush Report and Bush himself on the administrative provisions of S. 1850 too confusing (Chalkley 1951: 21). Still another commentator blames the "National Association of Manufacturers, and other lobby groups, [which] vigorously opposed any national science foundation." According to this commentator, NAM and others "got to the House committee to which the bill was referred and succeeded in smothering it" (Schriftgiesser 1951: 243). Indeed, NAM did meet with Bush aides on the Mills bill and with others in the Senate on a conservative alternative to 1850.[37]

Which of these specific factors explains the House Committee's actions is somewhat beside the point. We need to understand what made each of these factors possible. The answer lies in the organization of the American state and polity. Quick resolution of the dispute over S. 1850 was unlikely where interests were highly fragmented, where the state was divided, where the Congress was highly permeable, and where party discipline simply did not exist. With Democratic control of the legislature and the executive, party discipline would have assured passage of the compromise measure, if not of the original Kilgore bill, if Democratic representatives and senators had followed the call of their party leader, President Truman. Without party or executive discipline, there was no single position which could be articulated. In the absence of discipline, and with a legislature easily accessible to social interests, the Congress was torn by mixed messages from diverse interest groups, and well-positioned congressional conservatives were able to table the measure.

From Inner-Party to Inner-State Conflict

Elections in 1946 gave Republicans control of the House and Senate for the first time since 1928, thus creating a divided state in which Democrats controlled the executive (K. Jones 1975: 328). When the congressional session began in 1947, Kilgore was divested of his important chairmanship, and Senator H. Alexander Smith of the Senate Committee on Labor and Public Welfare—a New Jersey Republican who had close ties to Princeton University and was the senator behind Bush-inspired amendments to the Kilgore-Magnuson compromise in the previous congressional session—was given responsibility for science legislation (England 1982: 48; B. L. R. Smith 1990: 46). The improved structural position of the Bush forces—their improved capacity for informal influence of those with formal state power—delighted Bush and prompted Smith to brag to industry representatives that the changed situation would allow the Bush forces to "have the jump on the situation."[38] Smith asked Kilgore to cosponsor the legislation he intended to introduce, but Kilgore declined, telling Smith that his proposed legislation ran contrary to the basic principles that Kilgore believed should underlie science legislation.[39]

In preparing new legislation and, indeed, throughout the course of the 80th Congress, Senator Smith consulted Dr. Bush, Harvard president and deputy director of OSRD James Conant, and MIT president and OSRD leader Karl Compton.[40] Of the three, as with the past legislation, Bush was perhaps most deeply involved. In a letter to Conant at the beginning of the congressional session, Bush said, "there has been a lot of activity on the science bill lately and I have been up to my neck in a nice quiet little way."[41] As had Senator Magnuson, Smith received legislation drafting assistance and help in organizing hearings and floor debate from Bush assistant John Teeter.[42]

In addition to working with the OSRD inner circle, Smith sought the support of the military. He worked closely with U.S. Navy Secretary James Forrestal. Forrestal, like Smith, had close ties to Princeton, and Smith, on more than one occasion, sought Forrestal's advice on the proposed legislation. In one instance, Smith asked Forrestal's view on the inclusion of a military division in the science foundation, then told Forrestal that if the navy thought such a division inappropriate, Smith would "of course, want to give it very serious consideration."[43]

Smith was also concerned to obtain business support and consulted with industry representatives on the legislation. At the request of Smith, after the 1946 election, Randolph Major, research director of the New Jersey–based Merck and Company, arranged a dinner meeting at the University Club in New York. Included among the guests were members of the Directors of Industrial Research (DIR) and New Jersey's Senator Smith, as well as several other senators (Conner interview, 9/26/90).[44] Bush was invited, but did not attend. In a letter just prior to the meeting, Bush told Randolph Major: "I do not believe that I should definitely plan to join such a meeting as you are putting together unless and until I know that it is quite in accord with what is fitting and proper on my part, for I am still head of OSRD which is not yet dissolved."[45]

Smith found the dinner meeting quite productive and claimed that he had obtained the full support of the Research Directors on the legislation. In a letter to a senate colleague after the meeting, Smith bragged that he "had a 'field day' with the research directors of some of our most important research groups and sold them one hundred percent on our bill."[46] At the same time, in a letter to William Cole of Rutgers, Smith acknowledged the influence of the Research Directors. He told Cole, "I found them all unanimous in insisting that this [science] measure be limited to research in pure science, and not in applied science. . . . We actually took the words 'scientific development' out of the bill for this reason."[47] And although Smith believed fellowships and scholarships should be handled in different legislation, he acknowledged that they were included in the science foundation legislation at the insistence of George Merck and James Conant.[48]

On February 7, 1947, not long after the research directors' dinner,

Smith introduced his bill. This bill marked the second distinct legislative effort to create a national science foundation. The Smith bill, S. 526, called for *a large board* chosen by the president. The board was to select an executive committee who would in turn elect a director. In addition, though the proposed foundation was to include a national defense division, emphasis was placed on basic research. In keeping with Bush's views, the bill included no patent provision and excluded specific mention of the social sciences (Hodes 1982: 149–50; Rowan 1985: 129–30). Finally, unlike the amendments intended as a substitute to 1850 introduced by Smith in the previous session, the new Smith bill included recognition of the need for coordination in government research policy, and the bill included provision for an interdepartmental committee to *coordinate* federal research (K. Jones 1975: 330–31). Thus, a few of the philosophical underpinnings of Kilgore's legislative efforts survived.

On the same day that Smith introduced his bill, Senator Elbert Thomas introduced S. 525, a bill identical to S. 1850, the Kilgore-Magnuson compromise from the 79th Congress. Both bills were referred to Smith's Committee on Labor and Public Welfare. The bill reported out of committee on March 26th bore the number of the Smith bill and generally followed the contours of the original S. 526 (Jones 1975: 330). A number of efforts were made to amend the bill on the floor. Proponents of a provision guaranteeing equitable geographical distribution of resources were successful in their efforts to gain an amendment to the bill, but efforts to gain inclusion of the social sciences and a patent provision failed (333–34; McCune 1971: 121–5).

Most important, meetings between Truman, Budget Bureau officials, Bush, and other elite scientists, as well as Senate leaders, produced a compromise on the administrative provisions of the bill. The compromise, considerably weaker than the original Kilgore provisions, called for a presidentially appointed director who could be removed at the discretion of the president. The amendment embodying this compromise passed, but only, according to NSF historian Merton England, because Smith and possibly other amendment opponents misunderstood what they were voting for. Two amendments had been offered, one giving only the president the discretion to remove the director and another granting removal powers to the foundation board, as well as the president. Smith favored the latter amendment, but mistakenly voted for the former, giving it a one-vote margin of victory (England 1982: 76).

In different forms, the bill passed both houses and was sent on to the conference committee. The bill reported out of conference in late July differed considerably from S. 1850. First, while 1850 said nothing about the profession or interest board members were to represent, except to say they were not to be chosen on the basis of political affiliation, S. 526 said foundation members should be "eminent in the fields of the fundamental sciences, medical science, engineering, education, or public affairs," should have records of distinguished service, and should represent "the views of scientific leaders in all areas of the nation" (U.S. House 1947b: 1). Clearly, this was a coup for science elites. The board would not represent the views of all social interests, but of *scientific leaders.* Commissions established in the act to investigate specific scientific concerns (e.g., cancer) would be dominated by scientists. Each such commission would have six places for scientists and five for the general public (5). Also, the foundation was to be controlled largely by foundation members and the executive committee they elected—a clear snub to the administration. Finally, on the issue of administrative control, in a concession to nonelite scientists and educators, the bill suggested the president consult not only the National Academy of Science in making his board choices but also organizations such as the Association of Land Grant Colleges and Universities (2).

The bill took the foundation's policy role more seriously than did earlier legislation, but this role did not include consideration of applied science. The act directed the foundation to develop a *national policy* for basic research and education in the United States (U.S. House 1947b: 2). Even here the specifics of the agency's policy role were not clarified. A *coordinating role* was also given attention. The act created an Interdepartmental Committee on Science to be made up of departmental and agency representatives. The aim of the committee was to eliminate unnecessary duplication and to provide recommendations to the president, foundations, and other agencies in order to meet the requirements of the act.

The bill still provided the NSF a role in military and medical research, but explicitly excluded it from a role in atomic energy research. Specific fields of science for support were mentioned in the bill, but the social

Table 5.6 Second Legislative Round: Original Kilgore Proposals versus Final Smith Bill

	Original Kilgore Proposals	Conference Version (S.526)
Presidential appointment	Board and director	Board only
Coordination/ planning	Strong mandate	Vague mandate
Control/ administration	Business, labor, farmers, consumers	Probable domination by scientific elite
Research supported	Basic and applied	Basic research
Patent policy	Nonexclusive licensing	To be determined by foundation
Social sciences included	Not specifically mentioned	Not specifically mentioned
Geographical distribution of funds	No discussion	No discussion

sciences were not included. The patent issue was not skirted, but the foundation's patent policy was left up to the foundation to determine in line with the "public interest" (U.S. House 1947b: 6). Finally, the bill that came out of conference lacked the geographical distribution requirement of the amended Senate version.

Most important for the fate of the bill, the amendment giving the president the right to appoint the foundation director was eliminated in the conference committee (U.S. House 1947b: 3) (table 5.6). In this context, as with earlier legislation, the impact of the bulk of American scientists on the legislation was not significant. In the interest of creating a united scientific front on the science legislation, Cornell president Edmund Day founded the Inter-Society Committee for a National Science Foundation at the end of 1946 (England 1982: 65). This group was composed of over seventy scientific societies and included such scientific leaders as Harlow Shapley and Isaiah Bowman, opponents on the earlier science foundation effort.[49] Bush was not opposed to the committee, but according to NSF historian England, he kept his distance from it (1982: 65).

One issue on which member organizations of the Inter-Society Committee appeared to agree was that the science foundation should be controlled by *a single administrator.* Smith sought support of the committee, and, of course, a provision specifying that the foundation would be headed by a single presidentially appointed director was passed in the Senate, but it was eliminated in conference. This led one member of the committee to conclude that "the Congressmen most directly responsible for science legislation appeared to attach much less importance to the views of a two-thirds majority of scientists than they did to those of a few particularly prominent ones" (Wolfe 1947: 533).

Loss of the amendment providing for a single director appointed by and accountable to the president created division in the executive over whether President Truman should sign the bill. The military, in line with the previous support for Bush-inspired legislation, recommended signing, and the Public Health Service and other agencies also supported it. The Budget Bureau, which had strongly advocated an administrative structure that would be accountable to the president and not give part-time officials administrative responsibility, was divided: the leadership recommended a veto, but others felt it was not responsible to further delay development of a national research program (Rowan 1985: 143–46).

In the end, following the advice of the Budget Bureau leadership, in early August, Truman vetoed the measure. He said he did so with "deep regret" because he believed that in an effort to prevent the foundation from becoming dominated by politics, supporters of 526, if successful, would have created a government body "divorced from control by the people to an extent that implies a distinct lack of faith in the democratic process." In short, if he had signed the legislation, Truman believed the organization would have inappropriately undermined presidential prerogative and public accountability (Truman, quoted in K. Jones 1975: 340–41; see also England 1982: 81; McCune 1971: 133–34).

Truman's veto marks a second important delay point in NSF legislation. Failure of the first legislative effort to create a National Science Foundation was the product of the interaction of three factors: a permeable state, a divided executive, and a majority party—the Democrats—unable to enforce party discipline. In this second case, passage of a bill unacceptable to Truman can be attributed to Republican control of the legislature. However, it is by no means certain that Democratic control would have guaranteed a favorable vote on legislation acceptable to the president. Southern Democrats might have worked with Republicans to block bills to Truman's liking or crafted a measure—with or without the assistance of the science vanguard and business elites—unacceptable to the president.

In different senses, both failures were overdetermined. In the first case, any of the three structural factors I have pointed to—a permeable state, a divided executive, and an undisciplined majority party—could have doomed the bill, although in the first legislative effort to create an NSF they appear to have worked together. The second case—the case of the Truman veto—was overdetermined in the sense that, in the absence of a divided state (a Republican legislature and a Democratic executive), a plausible alternative scenario could have produced a similar outcome. For example, a coalition of southern Democrats and Republicans could have killed the bill in one of several committees.

At a general level, these two failures to pass NSF legislation can be attributed to the same cause: the structure of the American state and party system. The U.S. state and party system—then and now—has a greater capacity for veto than for policy innovation, and, as I will show, it took a structural modification to inhibit the state's veto capacity and permit passage of NSF legislation.

Another Go: Establishment of the National Science Foundation

Having failed to win approval for S. 526, at Truman's request, Senator Smith went to work on a piece of compromise legislation (McCune 1971: 143). Again, Smith asked for Bush's suggestions concerning the legislation, and Bush aide John Teeter was involved in drafting the bill. As he had done earlier, this time Smith also sought the advice of the military concerning establishment of a military division as part of the NSF. Charles Brown, a former OSRD staff member, later with the Office of Naval Research, was asked to prepare a possible amendment to keep military research outside the scope of the foundation. Finally, Smith sought the views of scientists and the executive concerning administration of the foundation.[50]

The new bill, S. 2385, was introduced in late March of 1948 with Kilgore as cosponsor.[51] This was a real piece of compromise legislation.

To overcome the threat of a presidential veto, the new bill called for the foundation director to be appointed by the president. Early versions left control largely in the hands of an executive committee, but this provision was eliminated from the final draft. Still, the legislation left the foundation to a significant degree *in control of scientists.* As with S. 526, the qualifications for board membership were eminence in science or public affairs, and all special commissions created by the foundation were to have a majority of scientists among their members. The bill did grant the foundation *a policymaking role,* but only in the areas of basic research and education. In what was apparently a slight coup for Kilgore, the bill included as part of the foundation's policy role assessment of the "impact of research upon industrial development and upon the general welfare" (U.S. Senate 1948: 6). In addition, commissions established by the foundation were to be responsible for providing surveys of the scientific fields they covered and, on the basis of these, to formulate and recommend an overall research program for that field.

Following the recommendations of the military, the bill called on the foundation to support defense research, but did not call for creation of a military division of the foundation. In addition, recognition of the concerns of nonelite universities is suggested in a provision that, while not establishing any specific distribution of resources, admonished the foundation to avoid "undue concentration" of support for research and education. Finally, like S. 526, the bill's patent clause left policy up to the foundation, and *the bill denied the NSF a role as lead agency for research and science policy* (U.S. Senate 1948).

S. 2385 was criticized by some scientists on the left for its failure to create a serious policymaking agency with coordinative capacity, for leaving control in the hands of private citizens, and for providing no guarantees of equitable geographic distribution.[52] The National Association of Manufacturers was internally divided about whether to support any science legislation at all, after earlier having supported the Bush plan.[53] Still, while not everyone was satisfied, the measure was broadly supported. It passed the Senate on May 5, 1948, and the House began hearings at the beginning of June.

After hearings, on June 4, 1948, the bill was favorably reported out of the House Committee on Interstate and Foreign Commerce (Chalkley 1951: 18). From the Commerce Committee, the bill went to the House

Rules Committee where it needed to be approved before it could be debated on the House floor (England 1982: 91). The Republican-controlled committee failed to report the bill out, and it died in committee.

One historian has argued that the Republican leadership may have purposefully blocked the bill because 1948 was a presidential election year and Republicans had "no desire to give Truman some twenty-five appointments to the NSF" (McCune 1971: 145). Another historian has suggested that opposition to the bill by a former Democratic Congressman, Fritz Lanham, and opposition from the National Infantile Paralysis Foundation helped kill the bill (England 1982: 91, 92).

Ultimately, however, the failure of this bill must be explained like the failure of the earlier bills in structural terms. A proximate cause is clearly a divided state. Republican determination to kill the bill might have been influenced by interest groups, and this would suggest a role for state permeability in the failure of this legislative effort. Again, however, given the structure of the American state and party system, one can imagine quite plausible alternative scenarios which would have led to the bill's demise. Whatever the proximate cause, the 1948 election did not turn out as Republicans hoped. Truman was reelected, and Democrats gained control of both houses, setting the stage for the final battle over the National Science Foundation.

Early in the 81st Congress, the same group of senators that had sponsored S. 2385 introduced S. 247, a bill essentially the same as 2385. By this time in early 1949, the National Association of Manufacturers had largely turned its attention to other matters.[54] Scientist organizations, including the Federation of American Scientists, urged a positive vote on the bill (*Bulletin of Atomic Scientists* 1949: 184), and Bush aide John Teeter continued to work for passage.[55] The bill passed the Senate rapidly and was sent to the House where hearings began at the end of March.

In mid-June 1949, the House Commerce Committee favorably reported out a bill that differed only slightly from the Senate version (Chalkley 1951: 18). Like its predecessor, the bill was reported to the House Rules Committee, where once again national science legislation got bottled up. A coalition of Republicans and southern Democrats—possibly at the urging of National Academy of Science President Frank Jewett—held the bill in committee, ostensibly out of fear of spiraling federal support for science (Rowan 1985: 180).[56]

The bill spent more than seven months locked up in the Rules Committee, finally being extricated on the basis of the "twenty-one day" rule (Chalkley 1951: 23). The rule, proposed by Rules Committee Chair Adolph Sabath, was an effort by a Congress dominated by Democratic liberals to mute the negative power—the power to stall and stop legislation—enjoyed by the Rules Committee, which was controlled by a conservative coalition of four Republicans and three southern Democrats. The rule was adopted by the House on a procedural vote in the 80th Congress, but was repealed by the conservative Republican–southern Democrat coalition in the 82nd after Democrats lost 29 seats (*Congressional Quarterly* 1982: 63).

While short lived, this rule created a temporary change in the structure of the state. By permitting the chair of any legislative committee that had reported a bill favorably and requested a special rule from the Rules Committee to bring the matter to the House floor if the Rules Committee failed to act within twenty-one days of the request, it limited the effects of state permeability, of decentralized power in the Congress, and of weak political parties. It prevented southern Democrats with seniority from working with Republicans to stop legislation in committee; similarly, it limited the effects of informal contacts between representatives of social interests and powerful legislators since these legislators could not kill bills in committee.

Once extricated from the Rules Committee, the bill passed the House on March 1 and, with few changes, was reported out of the Conference Committee at the end of April (Rowan 1985: 182). President Truman signed the National Science Foundation Act of 1950 on May 10th, nearly five years after the first National Science Foundation legislation was introduced and nearly eight years after Kilgore began to explore the possibilities of national research policy.

The final bill was a partial victory for Kilgore. The foundation's director would be selected by the president, not by the board he appointed. Nevertheless, the language of the bill virtually assured the organization would be *dominated by science elites*. Totally absent was any suggestion that the organization's board should represent a broad cross-section of social interests as Kilgore had suggested in his early legislation; indeed, the board was to be selected "to provide representation of the views of scientific leaders in all areas of the nation," and special commissions estab-

Table 5.7 Summary of Legislative Proposals and Final Legislation

	Populist Proposal (Harley Kilgore)	Scientist/Business Proposal (Vannevar Bush)	1950 National Science Foundation Act
Coordination/ planning	Strong mandate	Vague coordination mandate	Vague coordination mandate
Control/ administration	Business, labor, farmers, consumers	Scientists (and other experts)	Scientists (and other experts)
Research supported	Basic and applied	Basic only	Basic only
Patent policy	Nonexclusive licensing	Effectively no non-exclusive licensing	Effectively no non-exclusive licensing

lished by the foundation were to have "eminent scientists" in the majority (*Bulletin of Atomic Scientists* 1950b: 186–87).

Like S. 526, the National Science Foundation Act of 1950 required that the foundation make patent policy in "the public interest," but, unlike Kilgore's early efforts, the bill did not guarantee the granting of nonexclusive licenses on patents from government-funded research, nor did it provide any guarantee that patent rights would be assigned to the federal government, as Kilgore had originally wanted.

In terms of a central coordinating and policymaking role, as Daniel Kevles has suggested, the NSF established by the final legislation would be "only a puny partner" in the larger federal establishment (1987: 358). *The policy role the agency was given was vague.* It was directed to develop a national policy for basic research, not for all research, and the making of this policy was anticipated to involve the evaluation of existing government programs and the "correlation" of the foundation's research programs with those undertaken in the public and private sectors. The agency was not given responsibility for assessing the impact of research on public welfare as S. 2385 required and as Kilgore had clearly envisioned in his early legislation. Finally, the agency would undertake military research only after consultation with the Secretary of Defense, and although there was no provision in the act specifically denying the NSF a role as the central coordinating agency as there had been in earlier bills, the narrow focus of the legislation as well as the establishment of other government agencies in the years between 1945 and 1950 virtually assured that the agency would serve no such role (table 5.7).

The Twisted Trajectory of National Science Legislation

In this chapter, I have attempted to explain the protracted battle to establish the NSF in more than proximate terms. My analysis places proximate factors within the context of the structure of the U.S. state and polity. In many ways, the politics of research policy in the United States follows the pattern of social and economic policymaking. A range of comparative studies on economic and social policy suggest that centralized and more insulated states and programmatic parties are crucial in providing the basis for comprehensive and coordinated policies and that the absence of these characteristics explains the incomplete social provision in the United States and the state's inability to embark on coordinated and comprehensive industrial policy (cf. Shonfield 1965). Small wonder, then, that, given a fragmented state and parties with relatively weak discipline, we should have ended up, after prolonged debate, with a narrowly focused elite-dominated research policy agency and ultimately with a state further fragmented in terms of research policy responsibilities.

The case of the first episode in National Science Foundation legislation clearly illustrates the barriers state structure and party organization posed for rapid passage of science legislation. A highly permeable state with legislators vulnerable to constituent interests made a compromise necessary. Once established, the compromise was undermined when Wilbur Mills, a Democratic Congressman, broke ranks with his party and introduced a Bush-inspired bill, at the urging of Bush, a member of the administration, who, not incidentally, had the support of the military. Democrats controlled the executive and legislature, but the president was unable to control even the various branches of the executive, let alone members of his own party in Congress. Finally, the House committee to which the bill was sent was, in the absence of party discipline, easily accessible to social interests, and with the Congress torn by mixed messages from diverse interest groups, congressional conservatives—Republicans and Democrats—were able to table the measure.

In many parliamentary systems Kilgore's early bill would have passed without compromise, and the situation that existed in the case of the second legislative effort could never occur. But with a divided government, President Truman found it necessary to veto legislation that posed an explicit challenge to executive prerogative.

The third and the final efforts to pass national science legislation also illustrate problems of divided government, permeability, and nonprogrammatic political parties. In the second session of the 80th Congress, Republicans still controlled both houses and held the bill in committee to deny Truman a victory. The permeability of the Congress may have also been a factor, since some historians suggest and contemporaries believed that House committee members had been prompted to hold up the bill by nonstate actors (a former congressman and a medical foundation). In the final case, lack of party discipline and a committee system that gave power to a few conservative members, as well possibly as a situation in which a scientist (Frank Jewett) acting on his own could influence the committee, held the bill up for a considerable period of time, and it took a temporary change in the structure of the state which inhibited the effects of state permeability, decentralized power in the legislature, and weak party discipline to permit the passage of the 1950 act.

My analysis of the four efforts to create a National Science Foundation can be viewed as a comparison of four similar events across time. The first three cases are structurally similar. Each failure is explained by some combination of a permeable state, decentralized legislative authority, a divided executive, division between the executive and the legislative branches, and a Democratic party unable to enforce party discipline. In the final case, a House rule can be viewed as having changed the structure of the state or at least the effects of these structural factors. This change permitted passage of the act.

The tortured history of national science legislation brings to the fore several issues that have been of long-standing importance in efforts to understand American state building. First, as in other cases of state building in the United States and elsewhere, crisis—in this case World War II—became a watershed, providing a unique opportunity. The ability of state actors to utilize this opportunity was seriously constrained by structural limitations—a permeable and divided state, decentralized legislative power, and weak political parties—established earlier in building the American polity.

A second important and related issue concerns the creation of centralized state agencies. Kilgore's failure in this regard is consistent with other state-building efforts in the United States. As Amenta and Skocpol have noted, "economic mobilization for war meant centralization," but whereas World War II meant the establishment of permanent centralized agencies in some countries, in the United States centralized agencies were established on a temporary basis to provide coordination only during the war effort. Indeed, like several other U.S. war agencies, Kilgore's model of centralization and coordination, the OSRD, was created as part of the Office of Emergency Management. Not surprisingly, efforts at social and economic planning were also seen as temporary aberrations in time of war (Amenta and Skocpol 1988: 112). Thus, at some level, perhaps, the failure of Kilgore's boldest efforts was inscribed from the outset in the organizational history of the American state and polity.

Two final characteristics of this struggle are worthy of mention. First, it is important to recall that the implications of nonprogrammatic parties in the United States had a very distinctive character in the period under discussion. The Civil War left southern conservatism dominated by Democrats, not Republicans. These Democrats consistently opposed efforts at centralization and were generally opposed to increases in federal spending. Thus, the coalition of these Democrats with Republicans gave a particular character to this effort at state building.

Finally, efforts to create a National Science Foundation were driven by an issue that throughout U.S. history has created division between the legislative and executive branches: executive prerogative. In this case, the interest of elite scientists in controlling their own foundation meshed well with a history of congressional fear of ceding too much power to the president, and, indeed, presidential efforts to enhance the power of the president were certainly fresh in the minds of legislators. It was in 1938, after all, that Franklin D. Roosevelt proposed to reorganize the executive (Leuchtenburg 1963: 277). His efforts were rebuked by Congress.

Beyond highlighting issues of the organization of the American state, the case of the struggle to establish a national science foundation also raises important issues about two groups of social actors: business and scientists. My analysis of this episode illustrates that those who have studied this history have underplayed the importance of business interests in actually shaping policy and have been insufficiently attentive to the importance of the overlap between scientist elites and business. Recognizing the overlap between the groups is important because it suggests that business influence on policy may sometimes be indirect. Thus, while we do not generally think of Vannevar Bush as representing business interests, certainly his close ties with business leaders, as well as his participation in leading companies, shaped his thinking about science legislation in a manner that is apparent in *Science—The Endless Frontier.*

This study also highlights the importance of informal connections in the shaping of American politics. "Instrumental" Marxism has sometimes been dismissed as being too simple. Yet there is no indication that the particular organizational power of business allowed it to influence policy in the case under discussion. Rather, it was a dense web of overlapping social connections and interlocking directorates that extended directly into the state, but did not require businessmen at all locations in the state, which appear to have provided the basis for influence.

Skocpol has argued that "historically, America's relatively weak, decentralized, and fragmented state structure, combined with early democratization and the absence of a politically unified working class, has encouraged and allowed U.S. capitalists to splinter along narrow interest lines" (1985: 27). Given the structure of the American state, however, this has not inhibited business impact on policy. In the case of national science legislation, instead of an organized central body negotiating and cooperating with the state, business worked through informal channels and, in the end, business appears to have obtained what it wanted: a foundation that focused on basic research and was not likely to pose a threat to the traditional operation of the patent system. Indeed, given nonprogrammatic political parties and a fragmented and highly permeable state, business and other social interests—here scientists—are more likely to draw on informal contacts to shape policy than on formal organizational relations with the state. In addition, where state structure and the line between it and civil society is highly permeable, the formal role of social interests in influencing policy may come when they obtain positions as state managers (as Bush and others did) rather than through, for example, direct contacts between the state and peak associations.

Finally, this chapter in the history of American science policy points to the importance of the political and social role of scientists. Some political science literature from the 1960s and 1970s warned of the threat to democracy of the growing importance of science and scientists in American life (Lapp 1965; cf. Price 1965). Indeed, some literature stressed the growing political importance of scientists in the postwar period. Schooler, for example, focuses on "those *factors* that have made scientists influential over policy and policymaking" (1971: xiii). Little attention, however, has been paid to the role of scientists as an elite with interests they are trying to realize.[57]

But scientists involved in the debates over postwar research policy had two and possibly three distinct interests. First, of course, as Lapp has noted, "the elixir of federal funds [provided during the War] had proved tantalizing" (1965: 12). Both scientists directly involved in the war effort and those on the sidelines wanted to maintain high levels of federal support after the war. Second, although there was division among scientists, science elites, and certainly others as well, wanted to maintain control by scientists over the allocation of resources. Third, although there was some division in the scientific community, many physical scientists believed that the social sciences posed a threat to the resources available for other areas of science (Reingold 1987: 335).

The war and the debates over postwar research provided the context within which scientists could engage in what could broadly be termed a *collective advancement project.* Furthermore, the struggle can be seen, in part, as over who will make research policy. It was a sort of jurisdictional dispute (Abbott 1988) that pitted advocates of lay control against advocates of scientist control. Finally, this battle was largely won by scientists who gained control of the allocation of research resources through guarantees of presidential appointment to the NSF board and directorship.

Perhaps the ultimate establishment of the NSF should be seen as a partial victory for three distinct interests. First, state interests—executive interests—were realized in the provision that gave the president control over board and director appointment. Second, business successfully limited the foundation to a focus on basic research with no provisions posing a threat to the existing system of property rights embodied in patent law.

Third and finally, scientists gained control of what would years later become an important source of resources and established their right to peer control and professional autonomy. The real losers were those who believed in substantial democratic or popular control over research resources and those who believed that planning and coordination could be extended to the sphere of science policy.

6

FROM GRAND VISION TO PUNY PARTNER

Fragmentation and the U.S. Research Policy Mosaic

Nobody has—or could—come up with a readily comprehensible table of organization to explain the labyrinth of [scientific] agencies, foundations, consultantships, academies, and committees that has grown up in Washington in recent years.

Meg Greenfield, political reporter, 1963

It is clearly the view of members of the National Science Board that neither the NSF nor any other agency of Government should attempt to direct the course of scientific development and that such an attempt would fail.

National Science Foundation, Fourth Annual Report, 1954

Prior to 1945, Senator Harley Kilgore and his aide Herbert Schimmel had a grand vision—a vision of an organization that would link scientific research to the social and economic well-being of the nation, a lead central agency in the federal government for the promotion and coordination of basic and applied research. But as the debate over postwar research policy continued year after year, the responsibilities of the organization were progressively broken off and taken over by other agencies, and, as historian Daniel Kevles has said, the NSF became only a "puny partner"—one of several agencies—in the U.S. government science policy complex (1987: 358).

Historians agree that the five-year delay in establishing the National

Science Foundation left the United States with a fragmented or pluralistic system for federal funding of research and establishment of research priorities (Dupree 1972: 463; England 1970: 3; Maddox 1981: 269; Pursell 1966: 244–45; Rowan 1985: 201; U.S. House Task Force on Science Policy 1986a: 25, 30; Waterman 1960: 1341). Responsibilities originally intended to be fulfilled by a single science agency were undertaken by other agencies, including the Office of Naval Research (ONR) and the Research and Development Board (RDB) for military research, the Atomic Energy Commission for nuclear physics and related research, and the National Institutes of Health for medical research.

It is important to note that, once created, the National Science Foundation was established as the central, but not the only, supporter of basic research and was granted vague research policy and coordination responsibilities (Maddox 1981: 269; McCune 1971: 276; Rowan 1985: 206). Much existing research explains the delay in passage of the National Science Foundation Act and thus the fragmentation of the federal research policy system in proximate and ad hoc terms. Carroll Pursell, for example, asserts that "the perpetual realities of bureaucratic life and the new element of the East-West cold war sapped . . . [NSF of its] incipient primacy" (1966: 244). More generally, historians have concluded that other agencies "took up the burden" left when early NSF legislation did not pass (Dupree 1972: 463) or that these agencies "filled the void" (U.S. House Task Force on Science Policy 1986a: 26). The process in these explanations appears almost natural or automatic, not historically and organizationally conditioned.

The fragmented or pluralistic structure of research policymaking in the United States is the result of a protracted debate which—resulting from the organization of the state, state-society relations, and political parties—gave interests inside and outside the state time to break off little pieces of research policy into a range of agencies. The highly permeable state—a structure that facilitated delay—also allowed different interests to expand their institutional turf. Thus, the military created the Joint Research and Development Board to guarantee itself a role in postwar research policymaking. Similarly, the Public Health Service took advantage of the delay in the creation of the NSF to expand its own role in research policymaking (U.S. House Task Force on Science Policy 1986a: 29).

In this chapter I do three things. First, I flesh out the implications of the delay in establishing the NSF for the institutional contours of federal research policymaking in the immediate post–World War II period. Second, I consider the implications of the language of NSF's enabling legislation—itself clearly a product of the structure of the state and polity—for the early role the NSF played in the making of American research policy. Finally, I lend credibility to the account I have provided in this study by undertaking a cross-national comparative reconnaissance. I explore the relationship between national state and society organization and the character of national research policymaking.

Filling the Gaps in Military-Related Research and Development

Among the agencies created while the struggle for a National Science Foundation raged was the Atomic Energy Commission (AEC). Enabling legislation for this agency was introduced late in 1945 and was passed with bipartisan support in 1946. Perhaps the importance of such an agency in light of Nagasaki and Hiroshima, as well as the narrowness of its proposed mandate, allowed it to avoid the obstacles faced by NSF legislation. In the end, the organization centered on development of an atomic energy program under the aegis of a full-time civilian board appointed by and responsible to the president (Kevles 1987: 349–52).

One historian has argued that "with the setting up in 1946 of an independent Atomic Energy Commission, the still unborn [National Science] Foundation lost the most lucrative and dramatic field in all of science" (Pursell 1966: 244). But although Kilgore's staff had urged him to include atomic energy–related research in his proposal for a National Science Foundation, he declined (Jones 1975: 326). Indeed, inclusion of atomic energy research within the proposed science foundation was never a central issue of contention. Still, during the delay in passage of NSF legislation the AEC's responsibilities did expand into areas that had been considered likely areas of responsibility for the new foundation. In 1947, for example, Congress added $5 million to the Commission's budget, and the authorization was specifically earmarked for cancer research. In addition, the AEC gradually reevaluated its early decision not to support basic research that was not associated with weapons or reactors. In 1949 the commission initiated a program of basic research on a contract basis carried out by university scientists. In that year, the AEC financed research at 67 universities and listed some 144 unclassified research contracts at U.S. colleges and universities in areas ranging from biology and medical science to physics (Axt 1952: 96). In sum, prior to the creation of NSF, the commission had become the primary agency supporting nuclear physics research and an important funder of other basic research (Jones 1975: 357).

While the AEC did ultimately undertake a program of support for basic university research, the lead agency supporting academic research in the immediate postwar years was the Office of Naval Research (ONR) (Sapolsky 1979: 384; 1990). The idea for the ONR originated during the war among a group of young naval officers, mostly science Ph.D.s, known as the "Bird Dogs" because of their responsibility for "ferreting out problems in interorganizational relationships" (Sapolsky 1990: 9). Their idea was to create an office that would sponsor not just research by navy scientists but also work in civilian laboratories. The agency's aim would be to promote basic research for the long-run benefit of the navy's weapons' development capacity (Greenberg 1967: 134; Kevles 1987: 354; Sapolsky 1990: 44).

Navy Secretary Forrestal created the proposed office under temporary war powers in May 1945. The agency—the Office of Naval Research—was made permanent by an act of Congress in August 1946. The chief of the organization was given responsibility for "initiating, planning, promoting and coordinating the naval research program" (Powers 1947: 122). The ONR supported research in its own labs, but also in universities through contracts. ONR made an effort to maximize the flexibility of university contractors, and while navy officials determined the broad contours of research support, civilian scientists determined to whom awards would be made (Sapolsky 1990: 42–43). In addition to its support for university basic research, the ONR supported applied research related to naval missions and was established with a medical branch (Kevles 1987: 354–55; Powers 1947: 122–23; B. L. R. Smith 1990: 48).

The ONR played a major role in support of basic research for almost a decade after World War II, and received consistent support from Congress (Jones 1975: 359). Prior to the establishment of the NSF, the military was the "principal supporter of basic scientific research in academia via the Office of Naval Research (ONR), which moved quickly into the vacuum left by the absence of the NSF to build cyclotrons and betatrons, support the research of astronomers, chemists, physiologists, and botanists, and branch out into such unmilitary studies as meteors, the rare earths, and plant cells" (Kevles 1990: xv–xvi). Its early domination of American physics is made clear by the fact that in 1948 "nearly 80 percent of papers presented at the American Physical Society meetings were said to have been supported by Office of Naval Research money" (Kevles 1987: 363–64).

While congressional leaders and others were busy debating the inclusion of a military division in the NSF, ONR and AEC were happily consuming and controlling significant federal resources for research. At the same time, anticipating the elimination of OSRD, in June of 1946, the Secretaries of War and the Navy established the Joint Research and Development Board (JRDB). The board was intended to coordinate all research and development activities of interest to both departments and to promote a fully integrated research and development program for national defense. To facilitate coordination, the board was given "final authority to make allocation of responsibility for research and development programs between the Army and Navy" (Stewart 1948: 50). Following the model established by OSRD, the JRDB aimed to maintain the military's close relationship with civilian research and scientists (Powers 1947: 122; U.S. House Task Force on Science Policy 1986a: 32–33). The board used numerous advisory panels to guarantee direct participation by civilian scientists (Pursell 1966: 245).

In many ways the board was quite similar to OSRD, and not surprisingly Vannevar Bush was appointed the board's first chair. In 1947, the National Security Act created the Research and Development Board to replace the JRDB. Bush served as chair until October 1948, at which time he was replaced by Karl Compton (York and Geb 1977: 15). The new board employed a full-time staff of 250 augmented by 1,500 part-time consultants who served on the advisory panels that coordinated the thousands of military research and development projects (Forman 1987: 157).

By 1949, the Department of Defense and the Atomic Energy Commission together accounted for ninety-six percent of all federal support for university-based physical science research (Kevles 1987: 359). Contemporaries and historians have argued that delay in establishing the National Science Foundation led to the domination of American science by the military (Kevles 1987: 360; Rabinowitch 1946: 1). There can be no doubt as to the importance of the military in setting research agendas in American science (cf. Forman 1987), and certainly the delay in passage of legislation creating a central science policy agency did lead to the fragmentation of federal sources for research support, and produce "an irreparably stunted National Science Foundation" (Forman 1987: 183). But, as Forman notes, with or without an NSF, "cold war policy objectives guaranteed a prominent role for the military in supporting research" (1987: 226).[1]

Medical Research in the Absence of the National Science Foundation

The federal government's commitment to support basic medical research predates debates over the organization of postwar research policymaking by more than a decade. In 1930, the National Institute of Health (NIH) was created as part of the existing Public Health Service (PHS) with a mandate to support research on chronic diseases (Swain 1962: 1233). Despite the existence of the NIH, Bush and Kilgore hoped to include medical research in a national science foundation, and Bush's committee on medical research, as well as many medical administrators and researchers outside the government, called for a new independent medical research agency (Swain 1962: 1236; Bush [1945] 1960: 46–69). Lack of resolution in the science foundation controversy, however, created an opportunity for NIH/PHS officials to gradually shape a national medical science policy in a way that considerably expanded their own agency (Strickland 1972: 21; U.S. House Task Force in Science Policy 1986a: 29).

The process of developing a medical research program at the Public Health Service took place slowly. After the war, NIH hoped to have OSRD medical research contracts transferred to their agency, but Bush wanted the contracts to go to the proposed science foundation. It was not until it was clear that foundation legislation would not be quickly established that the OSRD contracts were transferred to the NIH. According to Donald Swain, these contracts "undergirded the emerging extramural research program" at NIH (1962: 1235–36). In addition, soon after the war the NIH began granting research fellowships, and in 1946 the agency created its own research grants office to coordinate its growing program in research support to investigators at universities throughout the nation

(1236). In 1949, PHS grant support for university and medical school research reached $8 million, with an addition of $7 million provided by the agency for construction of university research facilities. By 1950, the research figure alone was up to $11 million (Axt 1952: 102).

A National Cancer Institute was created by Congress under the Public Health Service in 1937 to support research on cancer in government laboratories, as well as in university labs. The Cancer Institute also supported advanced training for specialists and fellowships within the PHS (Dupree 1957: 365–66). The institute was "running smoothly," and there was widespread support among elected officials for government funding of medical research (Strickland 1972: 75; Swain 1962: 1236). In this environment, the Surgeon General recommended congressional authorization of several new disease-oriented institutes to be organized along lines similar to the Cancer Institute. In 1946, the National Institute of Mental Health was authorized, and in 1948 the National Heart Institute and the National Dental Research Institute were authorized (Swain 1962: 1236). In that same year, the National Institute of Health was renamed the National Institute*s* of Health, and the postwar medical research policy establishment was complete (U.S. House Task Force on Science Policy 1986a: 29). Thus, when the NSF was finally established in 1950, an existing agency was responsible for disease-oriented medical research, and NSF was left to focus more basically on "advancing our knowledge and understanding of biological and medical fields" (Bush 1960 [1945]: xii; Swain 1962: 1236).

Table 6.1 gives a brief summary of these major agencies.

The Early National Science Foundation

I have argued, and most commentators agree, that the lengthy delay in passage of national science legislation had a profound effect on the organization and nature of research policymaking in the United States in the postwar period. I have attempted to extend previous studies of postwar research policymaking, arguing that the organizational characteristics of the American State and polity play an important role in explaining the delay and thus the final configuration of postwar research policymaking. In addition to explaining the winnowed role the NSF ultimately undertook

Table 6.1 Federal Agencies with Research Policy Responsibilities Founded or Expanded between 1945 and 1950

Agency	Year	Research-Related Responsibilities
Atomic Energy Commission	1946	Support for basic research in atomic energy and other areas through contract
Office of Naval Research[a]	(1945) 1946	Support for basic and applied research loosely related to Navy's mission, contract and in-house
Joint Research and Development Board (Research and Development Board)[b]	1946 (1947)	Coordinate defense R & D
National Institutes of Health[c]	1946, 1948	External grants office to coordinate support for medical research; addition of new institutes

[a]The Office of Naval Research was created by the Secretary of the Navy under temporary war powers in 1945 and was made permanent by an act of Congress in 1946.

[b]The Joint Research and Development Board was created by the Secretaries of Defense and the Navy in 1946 and replaced by the Research and Development Board as a result of a 1947 act of Congress.

[c]The National Institute of Health was created in 1930, but did not become the National Institute*s* of Health until 1948 after several institutes had been added to the original organization.

in research policymaking and the fragmentation of the federal research policymaking system, the organizational characteristics of the U.S. state and polity have a central role in helping us understand the early National Science Foundation. In this section, I look at several issues that were left unresolved in the foundation's enabling legislation and which were confronted early in the organization's history.

A central issue animating the debate over a national science foundation was who should control it and how it should be administered. Early on, Kilgore proposed an agency administered by boards made up of a broad cross-section of social interests, including business, labor, consumers, and farmers. Kilgore gave up on this idea early in the debate, and a central issue of contention was whether the board of the foundation would select the director or whether the president should have that right. Underlying the broad issue of control and administration was the question of whether the foundation ought to be controlled by scientists or by representatives (ultimately the president) more broadly accountable to the public.

President Truman took his appointment prerogative seriously, and after the foundation's establishment, Truman considered nominating Frank P. Graham to be the foundation's director. Graham was a former history professor and president of the University of North Carolina. He had been an advisor to both Roosevelt and Truman and had a reputation of supporting the rights of workers and minorities (England 1982: 122).

Graham's proposed appointment was viewed as political by many slated for appointment to the first National Science Board, and the president avoided potential embarrassment by withdrawing the offer to Graham before it was ever made formally (England 1982: 123). The board members, themselves appointed by the president, demanded that the person appointed director be either a recognized scientist or science administrator. Ultimately the board recommended—and the president accepted the recommendation of—Alan T. Waterman. Waterman was a physicist and a protégé of Bush's good friend and colleague, MIT president Karl Compton. He had been a professor at Yale and had served on the OSRD staff, as well as having been the chief scientist of the Office of Naval Research (Kevles 1987: 359–60). In addition, Waterman was the long-time choice of Bush himself, and he was sworn in as director in April 1951, nearly a year after Truman signed the National Science Foundation Act (Wolfe 1957: 335).

The board itself represented the varied interests within the scientific community, but not proportionately, and went little beyond it. The National Science Board (NSB) was dominated by physical and natural scientists; only one social scientist was represented on it. Sixteen of the board members were university administrators, but in addition there were representatives of two prestigious foundations and among the two industry representatives was the president of General Electric. Although the board did represent a reasonable geographic cross-section, elite institutions had significant representation on the board, and indeed, Bush's long-time OSRD associate, Harvard President James Conant, was selected to serve as board chair (England 1982: 119; Rowan 1985: 189).

The interests that dominated the debate over the National Science Foundation held the balance of power in the agency and were in a position to shape its future. Industry retained a direct role in foundation governance, and of course, people like Conant, who had worked closely with Bush, were in a good position to understand the interests and concerns of business. In addition, the interests of the military, while not directly represented, were clearly understood by board members with wartime research experience, and both Waterman and his replacement as director had backgrounds in military research (Schaffter 1969: 31–32).

Fundamentally the board aimed to serve what they perceived as the interests of science. Waterman was certain that he and his colleagues were the "most capable of deciding what is best for the progress of science" (Waterman quoted in England 1970: 4–5), and he believed that science policy was a "matter primarily to be determined by scientists themselves" (Waterman 1960: 1342). Waterman's view was virtually identical to the stance taken by Bush and his science vanguard in advocating creation of a single science agency under scientist control and in justifying the need for the autonomy of Office of Scientific Research and Development during the war, and this philosophy of scientific organization was central to the wartime and early postwar collective advancement project of scientists.

With the board in place, the foundation was ready to tackle the issues left unresolved by the legislative battle. A central issue to be resolved was *the nature of the policymaking and coordinating role* the foundation was going to play. Kilgore had originally envisioned *a central science policy agency* that would coordinate research policy for the federal government and would be involved in establishing research priorities for the government and thereby the nation. Bush also wanted the science foundation to play some sort of policymaking role (Bush 1960 [1945]: viii).

By the time science legislation was signed into law, language giving the NSF a policymaking role was more narrow than Kilgore had wanted. Following the position of the Bush forces, the foundation was directed to develop a national policy for *basic* research, not for all research, and making this policy was anticipated to involve the evaluation of existing government programs and the "correlation" of the foundation's research programs with those undertaken in the public and private sectors. The agency was not given responsibility for assessing the impact of research on public welfare as S. 2385 required and as Kilgore had clearly envisioned in his early legislation.

President Truman was clearly committed to a policymaking and coordination role for the new foundation. In his budget message to Congress in 1952 he said, "the Foundation will formulate a broad national policy designed to assure that the scope and the quality of basic research in this country are adequate for national security and technological progress" (quoted in Wolfe 1957: 340; see also U.S. House Task Force on Science Policy 1986a: 27).

Delay in establishing the NSF permitted creation of institutionally entrenched interests. Truman was opposed from both within and outside the executive. Defense agencies opposed any NSF attempt to evaluate military research programs (Kevles 1987: 359), and other agencies opposed any coordination role for the new foundation that would threaten the existing division of bureaucratic turf (England 1982: 199). Allies of existing agencies inside and outside Congress also opposed any coordination role for NSF (Reingold 1987: 327). Based on his experience with government research, as well as his contact with university and government scientists, Waterman also opposed a serious coordination or policymaking role for the agency. He believed the foundation should focus on supporting research inappropriate for other agencies and should *not* be involved in evaluating the progress of other agencies (Sherwood 1968: 601–2).

In an environment in which implementing the provisions of the NSF act were left in the hands of an opponent of the broad policy role implied in the legislative history (an evaluation, coordination, and priority-setting role), the foundation was slow to take up its responsibilities. The foundation's first budget request was for some $8 million, and less than ten percent of these funds were to be allocated to the development of "a national research policy and for operating expenses." Only a meager $50,000 was to be set aside for policy development itself (Waterman 1951a: 251).

Some Budget Bureau officials continued to advocate a serious policymaking and coordination role for the NSF, despite clear opposition from Waterman and others inside and out of the administration (Sherwood 1968: 603–5). By 1953, President Eisenhower was faced with a large segment of the scientific community upset by calls for loyalty tests of scientists associated with the then-current Red Scare, small NSF appropriations, and distrust of basic research in the Defense Department. An advisor to the president urged Eisenhower to take some action that would mollify scientists. This created an opportunity for Budget Bureau supporters of a stronger NSF, when one of their number, William Carey, was given responsibility for drafting an executive order to reaffirm the administration's faith in scientific research and the NSF (606–7).

In drafting the order—actually two orders at the outset—Carey reasserted the need for the NSF to have strong coordination and evaluation roles, and through the order(s) he sought to establish the NSF as the government's central agency for basic research (Sherwood 1968: 607). The draft orders were opposed by agency administrators who feared loss of administrative turf and by university leaders who feared that a reorganization of federal research administration could cost them research dollars. Waterman, of course, flatly opposed centralizing government policy for basic research in the NSF (607).

Opposition led Carey and other executive office officials to redraft the order, and it was ultimately issued on March 17, 1954, as Executive Order 10521. The order called on the NSF to advise the president on the conduct of scientific research. The foundation was asked to determine where gaps and overlap in support for basic research existed, and other agencies were asked to consult with the NSF on basic research policy and to cooperate with it. Fundamentally, the executive order defined the foundation's policymaking and evaluation roles as cooperative, not regulatory (England 1982: 201–3). Ultimately, the order was a victory for those who believed that the NSF should not be a superagency for science, but a more limited body supporting basic research not already supported by mission agencies.

The responsibilities set forth in the order were considerably narrower than those envisioned by Kilgore and his allies. In the end, the foundation's policymaking role was reduced to "fact-finding" (Sherwood 1968: 610), and continuing dispute within the executive branch over the responsibilities of the NSF led to an agreement that called on the NSF to

> assume more leadership and to clarify the feasible goals in basic research. The Budget Bureau would make certain that other agencies consulted with the Foundation before submitting estimates for proposed basic research, and the Bureau, in consultation with the Foundation, would review agency action in order to strengthen research management. Most important, the Foundation would not be asked to assume *coordinating functions* which

> it was not equipped to handle or which would be resisted by other federal bureaus and scientists in general. (Sherwood 1968: 610–11; emphasis added)[2]

Two other issues that animated debate over creation of the NSF, but were not resolved through the language of the act, were decided in the early years of the foundation. First, while the social sciences were not named as a domain for support in NSF's enabling legislation, the act did permit the foundation to support areas of research not specifically named. This compromise provision permitted the foundation to begin supporting the social sciences in 1954 (Wolfe 1957: 337).

Scientifically underdeveloped regions of the country were not as fortunate as the social sciences when it came to the issue that concerned them: the final NSF legislation did not contain any requirement guaranteeing equitable geographic distribution of resources. The compromise measure instead called on the foundation to "avoid undue concentration" (*Bulletin of Atomic Scientists* 1950b: 186), and early administrators of the agency displayed no commitment to strengthening the "scientifically underdeveloped regions of the country . . . , in part because . . . they were strongly disposed to a best-science approach" (Kevles 1987: 365).

The highly permeable character of the state, fragmentation within the executive branch, and, indeed, the lack of discipline within the Democratic party made compromise on NSF legislation necessary. In many areas compromise took the form of vague language in the legislation itself. The indeterminate character of many of the provisions of the legislation made the appointment of a director especially important in determining the future of the foundation. Truman hoped to appoint a man with roots in the New and Fair Deals. It seems likely that Graham would have shared Kilgore's and, by extension, Truman's vision of the foundation, but the president was not sufficiently insulated to push through his first choice for director. Pressures from the first National Science Board led Truman to appoint Alan Waterman, the embodiment of the values of elites in the scientific community.

Waterman's appointment played an important role in shaping the NSF. He was committed to the "best science" philosophy of Bush and his colleagues. Waterman accepted the idea that the scientific community operated like a meritocracy: the "best science" should be funded and only scientists could judge what was best (Waterman 1960). This philosophy led to a disregard for geographic equity in distributing funds in the early days of the foundation and fundamentally shaped the nature of the foundation itself.

Waterman's appointment was also important for the future policymaking and coordination role of the foundation. He opposed an expansive vision of the policy and coordination responsibilities and had no intention of using NSF as the focal point for national science policymaking. There was opposition to Waterman's position within the executive, and indeed Budget Bureau officials attempted to force the NSF to take its policymaking role seriously.

Executive Order 10521 aimed to clarify the policymaking role of the foundation; it did little, however, to force Waterman and his board to engage in evaluation and coordination. Division within the administration and clear opposition from newly created and expanded science agencies made executive officials unable to enforce the Budget Bureau's vision of the NSF's policymaking role, and amidst the division the Eisenhower administration ultimately agreed to an extremely weak policymaking role for the foundation, a role Waterman and his colleagues could live with.

The Failure of a Grand Vision: A Comparative Reconnaissance

Throughout this study, I have viewed Kilgore's proposals for a research policy agency as a counterfactual case. The thread of my argument has been that if things had been different—most specifically if the state, political parties, and state-society relations had been organized differently—Kilgore's proposals might have become the basis for the federal system of making research policy after World War II. An in-depth cross-national comparison is beyond the scope of this study. I believe, however, that an initial comparative reconnaissance indicates the plausibility of the counterfactual underpinnings of my argument. My intention here is to bolster my counterfactual case by showing that, under conditions that contrast markedly with those of the United States, other countries have developed systems of research policy that share a lot in common with Kilgore's early proposals. In this context, it is worth considering the comments of Stanley Lieberson. Lieberson argues that "we can have maximum confidence in a counterfactual inference only when it is based on a high level of confidence that we know what would have otherwise occurred" (1992: 13). By providing comparative data, I am attempting to raise confidence about what might have occurred in the United States had conditions been "otherwise."

Thus, in this section I consider, on the one hand, existing research in a comparative frame, which clearly illustrates the role played by the structure of the state and civil society in shaping policy outcomes. From these cases it is clear that policy planning and coordination capacity—two components of Kilgore's early proposals—are associated with states that are not divided, permeable, or fragmented like the U.S. state and societies that are also more organized and centralized than is the case in the United States. Equally, several of these cases suggest that in policy areas outside of research, coordination, and planning are or were inhibited in the United States by the structure of the state and civil society.

On the other hand, I illustrate a correlation between state structure and national research policymaking characteristics; Kilgore-type research policymaking is associated with states and societies organized differently than those in the United States across several important dimensions. In addition, by showing that a number of elements of Kilgore's proposals are found in the research policymaking systems of other countries, I lend support to the claim that the Kilgore proposal was intrinsically viable or at least not intrinsically unviable—that is, that democratic control, centralization of policymaking, coordination, and planning are not impossible in the research policymaking area.

There are several policy domains in which one can see that the organization of state and civil society had effects similar to those I attribute to these factors in shaping the outcome of the struggle over establishment of a federal system for research policymaking in the United States in the postwar period. I begin with the case of industrial policy.

In his research, Peter Hall (1986) found that state structure and the organization of social interests was central in determining policymaking capacity. Hall explores the process of economic policy formation in France, Great Britain, and to a lesser degree, West Germany in the post–World War II period (table 6.2). In the area of state-directed industrial policy, Hall found that France was the most successful of his three countries in developing and sustaining a coherent and selective or tar-

Table 6.2 State-Society Organization and Industrial Policy Coherence

Country	State	Society	Policy Coherence
Great Britain	Weak coordinating capacity	No capacity for consensus building	Not coherent
France	Centralized, strong coordinating capacity		Coherent/state organized
Germany		Strong bank/firm coordination	Coherent/private-sector organized

Source: Hall, 1986.
Note: In the interest of providing tabular summaries of the relationship between state and society organization and policy outcomes, tables 6.2 through 6.6 simplify the organizational distinctions made in the text of this and previous chapters. Thus, in this table incoherence in the case of Great Britain means the country lacks capacity for sectoral planning, while coherence in the case of Germany refers to the capacity, not of the government but of the private sector, to promote industrial restructuring. To take another example, in table 6.3 low state capacity in the case of Great Britain means civil service opposition to public works programs, while low capacity in the case of the United States means lack of an institutional mechanism for channeling expert opinion and a fragmented system of policymaking.

geted policy (a policy that, in my terms, requires coordination and planning). He attributes this success to the structure of the state and the organization of capital. The French state is highly centralized. Training of state bureaucrats leads to a strong esprit des corps, and the Ministry of Finance has clear responsibility for monetary, fiscal, and industrial policy. The organization of the state has given it the capacity to coordinate economic policy areas and make selective interventions into industrial sectors. In addition, coherent industrial policy has been facilitated by industrial capital's dependence on finance capital and by the dependence of banks on the state.

In Britain, although there has been a concerted effort to utilize industrial policy in the postwar period, efforts have been hampered by a need to rely on consensus between business, labor, and the state. In addition, the British state lacks the capacity for true sectoral planning. Finally, in West Germany, state-led industrial policy has been relatively unimportant. Instead, a powerful financial community has played an important role in industrial organization through close links between firms. Banks' equity holdings in firms and government grants have led to sectoral rationalization.

Table 6.3 Foreign Economic Policymaking

Country	State	State-Society (Differentiation)	Society	Policymaking Capacity
United States	Decentralized	Low	Decentralized	Low
Japan	Centralized	Low	Centralized	High

Source: Katzenstein, 1978.

Peter Katzenstein (1978) examines foreign economic policymaking capacity. In contrasting the United States and Japan (table 6.3) Katzenstein argues that national foreign economic policy objectives are shaped by the "ideological outlook and material interests of the *ruling coalition*" (1978: 306). But the ability to carry out these objectives depends on available policy instruments, and Katzenstein contends that "the number and range of policy instruments emerge from the differentiation of state from society and the centralization within each" (308). In looking at the United States, Katzenstein highlights federalism, the separation of powers, the seniority system in Congress, and the absence of party discipline as fundamental characteristics of the state. Business, too, is characterized by organizational decentralization in the United States. This structure makes successful conduct of foreign economic policy difficult and leaves policymakers with "few instruments of limited range to pursue their objectives" (311).

By contrast, Japan has a powerful centralized state and civil society, and the degree of differentiation between them is low. Japan's ruling party monopolized control of the state for over three decades and maintained powerful discipline in the state bureaucracy. This structure and close relations between big business and the state, Katzenstein suggests, facilitate consistent policy (or, in other words, coordination and planning), using a large number of policy instruments that permit direct intervention in specific sectors and firms (1978: 297, 313–16).

Providing a less detailed historical analysis than the previously mentioned studies, Wilensky and Turner (1987) come to similar conclusions. They compare industrial, labor market, incomes, and social policies in eight first-world democratic capitalist countries and conclude that successful implementation of these policies depends on elite awareness of the

Table 6.4 Economic and Social Policy Coordination

Country	State	Society	Policy Coherence
United States	Decentralized	Decentralized	Low
Sweden	Centralized	Centralized	High

Source: Wilensky and Turner, 1987.

interdependence of these policy spheres and the *existence of national bargaining structures* that enable policymakers to act (1987: 1). Successful bargaining structures generally are based on strongly organized and usually centralized interest groups, especially labor and capital (10). Such structures provide avenues for elite influence on policy (11).

According to Wilensky and Turner, in comparison with countries like Sweden, the United States is unsuccessful in its efforts to create coherent (or coordinated) social and economic policies because its political economy is highly fragmented and decentralized (table 6.4). Interest groups are not "constrained by the necessity of national bargaining and tradeoffs [and] are in a position to act out their most parochial striving, reinforcing an already advanced state of paralysis" (1987: 15). In contrasting the United States' policy failures with successes in other countries, Wilensky and Turner point to a range of specific structural factors to explain U.S. shortcomings. These factors include "the division of powers between the executive branch and Congress; the incapacity of administrative agencies to link trade policy and ALMP [active labor market policy]; the complexities of federal-state relations; the lack of institutionalized channels through which the main actors can resolve conflict, feed back intelligence, and participate in policy, implementation, and outreach" (23).

Similar factors are used by analysts to explain the late and uneven development of social welfare provision in the United States. Orloff and Skocpol explain the development of expansive social welfare provision in Britain in the early twentieth century in terms of the existence of an insulated civil service, programmatically competing political parties, and a legacy of centralized welfare administration (table 6.5). The British civil service was reformed in the 1870s. Reform of the civil service prevented political parties from relying on patronage to satisfy constituents. Conse-

Table 6.5 Social Welfare Policy

Country	State	Political Parties	Policy Legacies	Welfare Provision
United States	Fragmented, permeable, low bureaucratic capacity	Not programmatic, patronage		Uneven, limited
Britain	Insulated civil service	Programmatic, disciplined	Centralized welfare administration	Expansive

Sources: Orloff and Skocpol, 1984 and Orloff, 1988.

quently, British political parties turned to formulated programs "to appeal through activists to blocks of voters" (1984: 740). Having appealed to voters and persuaded powerful constituencies of the need for a range of welfare benefits, the discipline of the British Liberal Party made passage of necessary legislation a relatively simple matter (741).

By contrast, the United States lacked an established civil bureaucracy, and at the time, Progressive political reformers were struggling against the corruption of patronage politics. Parties depended on patronage to attract constituents. They were not programmatic parties looking for programs to attract organized constituencies (Orloff and Skocpol 1984: 742). In this context, people "doubted that social spending measures could be implemented honestly" (743). In some states, purely regulatory reforms calling for little increase in discretionary spending were able to pass. In addition, workmen's compensation and mother's pensions, neither of which mandated new fiscal functions, were implemented in many states during the early twentieth century (745).

In her own work, Orloff extends this analysis of early efforts at social welfare provision and argues that in contrast to many European nations, the United States developed a system of social provision which is incomplete and lacks systematic connections as a result of the "ways in which U.S. state formation shaped the American political universe" (1988: 41). During the New Deal era, although ultimately the Social Security Act was passed, American social rights "were compromised by uneven national standards, the sharp distinction between assistance and insurance, and the omissions of health insurance, employment assurance, and allowances for all needy children" (79). As Orloff shows, this "largely reflected the inability of the New Deal social reform coalition, led by the Roosevelt administration, to overcome the deep resistance of congressional conservatives and some congressional constituencies to the changes they wanted to effect in American social policy" (79). Roosevelt's failure occurred within a context of division between the executive and the legislature, and despite Democratic control of the legislature, lack of a means to enforce party discipline allowed legislative conservatives from the President's own party to abandon him and side with their Republican colleagues.

These analyses cover range of policy domains from industrial and fiscal policy to foreign economic and social welfare policy. All indicate the centrality of some combination of the organization of the state, political parties, and social interests in determining the state's capacity to implement policy. While the degree of emphasis varies, all the studies highlighted here suggest that one must look at the fragmented character of the state—divisions between the executive and the legislative branches of government and divisions within each—and the nonprogrammatic and nondiscipline-enforcing character of American political parties to understand policy outcomes. Several also highlight the organization of social interests in determining policy outcomes.

Not only do these studies show the importance of state and society factors in determining policy outcomes, but also they suggest that across several policy domains these structures affected policy outcomes in the United States in a very particular way. In one way or another, all these studies suggest that nonprogrammatic political parties and a fragmented and permeable state make policy implementation difficult, limit the policy instruments available for use by policymakers, often delay policy implementation, and frequently prevent policy coordination.

While the case of the struggle over National Science Foundation legislation is distinctive in many ways, it also bears similarities to the accounts provided by the analysts I have discussed. As I have argued throughout this study, at different conjunctures state structure, party structure, and state-society relations led to the narrowing of proposals for a National Science Foundation, led to a delay in passage of legislation, and inhibited policy coordination, directly through fragmentation and indirectly through promoting delay in passing legislation.

The existing comparative research that I have explored indicates that the state, political parties, and state-society relations and organization are important causal factors or independent variables in determining policy outcomes. What of the dependent variable: policy outcomes? One could argue that Kilgore's proposals were never viable, that they were inherently unworkable. While again it is impossible to definitively resolve this dilemma, a review of the organization of science policy in several countries suggests that many elements of Kilgore's proposals are in fact workable and, indeed, since World War II have been implemented in several countries. In addition, while it is impossible in this brief exploration to definitively attribute causal efficacy to the independent variables I have highlighted in the U.S. case, my review here will at least indicate the existence of a correlation between state, party, and society organization and the nature of science policy in several national cases.

Taken together, Kilgore's early proposals have several elements that were not included in the final NSF legislation and against which it is possible to compare other national research or, more broadly speaking, science policymaking institutions. First, Kilgore proposed a single *central* agency, rather than a set of agencies, which would be responsible for making science and technology policy. Second, Kilgore's legislative proposals stressed the importance of *coordinating* different elements of science and technology policy and, indeed, coordinating efforts with the private sector. Third, Kilgore stressed the importance of *planning.* While this element of his proposal is nowhere elaborated, serious priority setting in line with national interests and a long-term, as against a short-term, policy orientation is perfectly compatible with Kilgore's New Deal roots. Finally, one of Kilgore's early bills suggested the importance of institutionalizing a role for important *social interests* in making research policy. Among other interests, Kilgore's legislation specifically named scientists, workers, business, and farmers.

In a cross-national comparative study of science and technology policy in France, Germany, the United Kingdom, Sweden and Japan, Leonard Lederman notes that "there is general agreement that the U.S. R&D system and organization are at the pluralistic, less centralized, and market-

Table 6.6 State-Society Organization and Science and Technology Policy

Country	Degree of Centralization		Characteristics of Science and Technology Policymaking			
	State	Society	Centralization	Coordination	Planning	Social Interest Role
France	High		High	Yes	Yes	
Japan	High	High	Medium	Yes	Yes	Yes
Sweden	High	High	Low	Yes	Yes	Yes
United States	Low	Low	Low	No[a]	No	No

[a]Since the Kennedy administration there have been several efforts to improve coordination in research policymaking.

oriented end of the spectrum; the French system and organization are at the more centralized, planned, and strategically targeted end of the spectrum; and the U.K., the FRG, Sweden, and Japan are somewhere in between" (1987: 1128) (table 6.6). This is quite a sweeping statement, but as a first cut it is helpful. Actually, none of these countries has one single agency responsible for all science and technology policymaking. On the other hand, at the more centralized end of the spectrum, several of these countries focus science and technology policy in only a very few agencies, and several have one or more prominent agencies with considerably more responsibility and power than the NSF has ever had or was intended to have in the legislation that finally passed.

In terms of the level of fragmentation of science and technology policymaking, the Japanese system contrasts sharply with the system that emerged in the United States after World War II. Three agencies have primary responsibility for science and technology policy and R&D funding in Japan. They are the Ministry of Education, the Science and Technology Agency (STA), and the Ministry of International Trade and Industry (MITI) (Cheney and Grimes 1991: 7; Ronayne 1984: 210). The Education Ministry is responsible for supporting basic and applied research in Japanese universities. MITI supports work in its own laboratories, as well as providing loans and equity support for risky research with commercial applications. The Science and Technology Agency, along with Japan's Council for Science and Technology, is responsible for coordinating science and technology policy.

In France, like Japan, research policy is considerably less fragmented than it is in the United States. Two agencies are primarily responsible for science and technology policy: the Ministry for Research and Higher Education coordinates university research and, through the National Center for Scientific Research, supports basic research. The Ministry of Industry, Telecommunications, and Tourism serves the same function for industry-related research (Lederman 1987: 1128).

In the area of coordination and planning there are also several countries that have greater capacity than the National Science Foundation, or for that matter any other U.S. federal agency, has ever had. In Japan, as I have noted, the Science and Technology Agency is the primary body responsible for coordinating national science and technology efforts. The STA has its roots in the Scientific and Technical Administration Committee created in 1948 as the nation's coordinating body for R&D. In 1956, the committee was given additional responsibility and transformed into the Science and Technology Agency (Ronayne 1984: 213). The STA is headed by a minister of cabinet rank and compiles and submits Japan's national science and technology budget. The STA also coordinates the research and development of all ministries and state agencies with the exception of MITI. As well, the agency assists in the establishment of long-term science and technology priorities and promotes national projects in new fields "requiring quick and energetic exploration" (Roynane 1984: 213; see also Cheney and Grimes 1991: 9).

In France, the coordination and planning of research and development (policy) has been the order of the day since creation of the Office National des Recherches Scientifiques et Industrielles et des Inventions in 1922 (National Research Council 1940: 195). This office's purpose was to "provoke, coordinate and encourage" all types of research, especially research with industrial importance (Gilpin 1968: 156). In 1933, at the behest of socialist-inclined scientists, a Superior Council of Scientific Research was added to coordinate basic research (158). Then in 1958, an Interministerial Committee for Scientific and Technical Research was established to serve as the science and technology coordinating body in the French national government. The committee recommends all research policy and is responsible for developing the science and technology section of the national five-year plan, which establishes national research and development priorities (Gilpin 1975: 116; Lederman et al. 1986: 73). The existence of a science budget in France gives the government an opportunity to obtain an overall view of support for science and, on the basis of this view, to reorient resource allocation.

While Sweden's research policymaking system is more fragmented than is the case for France and Japan, the national government does promote coordination and planning in science and technology policy. Since the 1940s, several agencies have funded research in Sweden (Dorfer 1975: 172; Lederman 1985: 138). In 1962 the Science Advisory Council was created to coordinate science policy and integrate science policy with other policy spheres. The Ministries of Industry and Education also play a role in coordinating research policy through nationally oriented sectoral policies (Dorfer 1975: 174–79). In 1979, coordination was improved with creation of a Council for Planning and Coordination of Research (Lederman 1987: 1129). This council periodically formulates national science plans.

Finally, while the United States lacks a mechanism institutionalizing the input of social interests into the science and technology policymaking process in a way that parallels Kilgore's early proposal, Japan, at least, has such mechanisms. Japan's foremost body for social interest advice on research policy is the prime minister's Council for Science and Technology. Council members include ministers, senior educators, industrial managers and scientists and engineers (Lederman 1987: 1131). MITI's Industry Structure Council makes recommendations that form the basis for the future orientation of national R&D. The council includes thirty members who represent industry, labor, and universities (Ronayne 1984: 214). In addition, MITI employs a range of advisory councils and consults with industry associations to ensure that government-conducted and -sponsored research meets the interests of the private sector (Lederman 1987: 1131).

Of course, only detailed historical work can substantiate the causal character of the relationships between state, party, and society structures and the organization of research policymaking. In any case, there are important methodological questions about historical research that employs a case-control approach. Concerns have been raised about the possibility of meeting equivalence and independence criteria in the comparison of national cases (Sewell forthcoming: 22–25). The existing literature suggests that U.S. officials may have played a role in defining the organization of postwar research policy in Japan during U.S. occupation of that country, and so the U.S. and Japanese cases probably cannot be considered independent (Batholomew 1989: 278). Nevertheless, at the very least, there is a clear correlation between the structure of states, political parties, and social interests and the organization of research policymaking in the postwar period. As the counterfactual that ran through my account of efforts to establish a postwar research policy agency suggested, and as this comparative reconnaissance confirms, countries with centralized and relatively impermeable states and programmatic and disciplined political parties tend to have research policymaking organizations with characteristics that have much more in common with Kilgore's proposals than with the system for research policymaking created in the United States after World War II.

In Japan and France, states are centralized and parties are considerably more programmatic and disciplined than in the United States, and certainly Japan and Sweden have systems of national bargaining, as characterized by Wilensky and Turner (1987). Japan and France have relatively centralized systems for research policymaking. In all three countries, coordination and planning play important roles in research policymaking. In Japan a broad range of social interests have an institutionally guaranteed voice in research policymaking. In Sweden, while much science policy institutionalizes the voices of scientists and civil servants, in the postwar period sectoral policies—which include R&D components—have always included input from capital and labor peak associations (Martin 1985).

Conclusion

Earlier, in chapter five, I argued that the delay in passing legislation to create the National Science Foundation can be explained in terms of the organization of the state and state-society relations and the character of political parties. The permeability of the state and the lack of discipline in the Democratic party and in the executive branch caused National Science Foundation legislation to be held up for five years, or eight years if one begins with Kilgore's earliest proposals for a national science and technology agency.

Passage of NSF legislation was not the end of the struggle over the shape of a national science policy agency. There was, as I have shown in this chapter, substantial institutional fallout from the delay in establishing a National Science Foundation. Already fragmented interests used the delay in passing NSF legislation to create several new agencies and to argue for substantial budgets to carry out their legislated or otherwise specified mandates, in the absence of a central science policy agency. Thus, while delay in passing NSF legislation is the proximate cause for the proliferation of research policy agencies in the period after World War II, the cause of the substantial delay—thus the indirect cause of the proliferation—is the organizational structure of the state, state-society relations, and political parties. A permeable state permitted a range of social actors to influence national science legislation, and a coalition of southern Democrats and Republicans was able to stall the legislation in committee on several occasions. Divisions between the executive and the legislature were also responsible for one specific delay. In some sense, we must conclude that a fragmented state begot further fragmentation.

The permeability of the state—which gave social interests easy access to congressional committees—and the inability of the Democratic party to enforce adherence to any program meant the language of the final legislation was vague and left implementation up to the Truman and Eisenhower administrations. But at this point, the administration confronted opposition from newly established or strengthened science agencies with their own interests and patrons. When the NSF was finally established, the interests behind several independent science agencies were bolstered by their new institutionalized security, and they opposed a broad role for the National Science Foundation. Truman also faced opposition on his choice of foundation director, and the director selected had no intention of creating a strong central science policy agency. The vague language in the NSF legislation and the state structure (in the form of several science policy agencies and division within the state and outside the state, which led Truman to appoint Waterman) led to a narrow reading of that language.

While the details of this account might be challenged, the broad contours are clearly bolstered by the comparative analysis I undertook in this chapter. Fragmented and permeable states and nonprogrammatic political parties have histories of inability to implement policies, of delayed policy implementation, and of difficulty in coordinating policy. Broadly as well, it is clear that more centralized states with programmatic political parties and more highly organized social interests have systems for making science and technology policy that bear more resemblance to Kilgore's proposals than they do to the institutional configuration that emerged in the United States after World War II. In short, the evidence suggests a correlation between state-society organization and the contours of national science and technology policymaking.

7

POSSIBILITIES AND PROSPECTS

Research Policy at a New Institutional Divide

> Unlike other countries, we have not developed coherent national science policies. Indeed, the very idea is abhorrent to many. Our free enterprise laissez-faire system has served us well during periods of expansion and growth, but in retrenchment the development of more formal science and technology policies seems essential if we are to preserve the best aspects of our system.
>
> Allan Bromley, as president of the American Association for the Advancement of Science, 1982[1]

> The United States Government currently lacks an adequate mechanism for determining which R&D programs should have national priority. This is one reason the U.S. is not as strong as it should be in competing in global markets.
>
> Technology Policy Task Force, Committee on Science, Space and Technology, U.S. House of Representatives, 1989

> When it comes to technology, U.S. public policy can no longer afford to be preoccupied with basic research and military issues; economic security and industrial competitiveness are also vital considerations.
>
> B. R. Inman and Daniel F. Burton, member and executive vice president, Council on Competitiveness, 1991

I have argued throughout this study that historical junctures of various sorts widely viewed as national crises create opportunities for changes in policymaking institutions and policy. In the case of the genesis of the research policymaking system in the United States, World War II was such a crisis. But while crises make change possible, the nature of change, as I have shown for the case of the National Science Foundation and the federal research policymaking system after the war, is shaped in fundamental ways by the organization of the state and society and the discourses alive at the time.

State-society organization has not changed much over the past forty years. The permeable and fragmented character of the state and the nonprogrammatic, undisciplined character of American political parties still means that *coordination and planning*—of the type Kilgore hoped to achieve through creation of a centralized research policymaking agency—is unlikely and further fragmentation is likely. But the difference in the nature of the current crisis from the wartime crisis means a change of discourse, as well as a transformation of group interests and possible alliances. In the face of fiscal belt tightening and economic hard times, the ability of academic scientists to transform their still formidable symbolic capital (cf. Kleinman and Kloppenburg 1991) into policy and state organizational change is declining. By contrast, the (rhetoric of) economic crisis is bolstering the symbolic capital of high-tech industries, and while state-society organization is likely to constrain their ability to promote research and technology policy coordination and planning, they are in a position to push priorities away from basic research toward the development of generic technology and precompetitive R&D. In turn, scientists may increasingly need to link their funding futures to the objective of high-tech industry and away from arguing that basic research must be funded, as it is the bedrock of national economic well-being.

In this chapter, I explore recent research and technology policy debates and the future of such policy. In the first section, I provide some important historical background on a central crisis intervening between World War II and the current crisis of competitiveness: the Cold War. The Cold War led to a continuation of state building for research policymaking, but the agencies created furthered the fragmentation of research policymaking and did little to improve coordination and planning.[2]

In the second section of the chapter I discuss two prominent proposals made in the late 1980s. The failure of one points to the continuing constraints imposed by the fractured and atomistic character of federal policymaking in efforts to increase policy coordination and planning, and the second suggests that the Kilgore legacy is still alive, but is not likely to lead to significant policymaking innovation.

In the final section of this chapter, I explore the current policymaking environment. I elaborate the argument I pointed to above concerning the reconfiguration of interests and alliances, and I explore what this reconfiguration is likely to mean for the future of technology policy in the context of a fragmented state with little capacity for policy planning and coordination.

My discussion of contemporary debates shifts focus from research policy narrowly defined to the broader category of technology policy. This shift is justified by changes in the terms of policy discussion that have occurred in recent years. While Kilgore articulated a program that is clearly a precursor to technology policy proposals of the 1980s and 1990s in its aim to promote transitions between fundamental scientific research, technology development, and commercialization, the elite science discourse that shaped policy from World War II through the end of the Cold War implied that U.S. economic well-being depended only on government funding of fundamental research. Federal support of fundamental research would lead to discoveries that would automatically be translated into useful technologies. This elite science discourse has begun to lose its ground, and there is now growing consensus that some type of active and more comprehensive technology policy is necessary if the United States is to be(come) competitive with its European and Asian trading partners.

Organizational Innovation and the Cold War

The Cold War bolstered the credibility of research as a source of protection against ostensibly hostile foreign nations. Driven by national embarrassment and worry, Cold War efforts led to further fragmentation of research policymaking and did little to improve the coordination of research policymaking.

On the heels of World War II peace, the Cold War between the United States and the Soviet Union was a central force in shaping the contours of the federal research policymaking system in the United States in the late 1950s and the early 1960s. The launch of the Soviet satellite, Sputnik, in

October 1957 played a particularly important symbolic role in prompting the creation of several new federal organizations. In light of Sputnik, government officials called for development of a U.S. space program in the interest of national security (McDougall 1985: 97). Scientists, including Vannevar Bush, called for increases in federal support for scientific research, also in the interest of national security (159).

In the context of national hysteria over the Soviet lead in space technology, the National Aeronautic and Space Administration (NASA) was created a year after the launch of Sputnik, in October 1958. NASA was the product of a bill proposed by the Eisenhower administration. The agency was to be administered by a single director and was to split responsibility for space programs with the military. The Defense Department was to restrict its efforts to secret programs and programs with direct military application (McDougall 1985: 172, 176).

A second important agency formed in the wake of the launching of Sputnik was the Defense Department's Advanced Research Projects Agency (ARPA).[3] The idea for ARPA is attributed to James Killian, President Eisenhower's science advisor, and Neil McElroy, Eisenhower's Secretary of Defense. According to Defense Department documents, the agency was in part a response to the military need for high-level attention to promising long-range research projects (U.S. House Task Force on Science Policy, 1986b: 51). The agency's mission was to explore advanced technology.

Since its creation, ARPA has contributed to important technological innovations in areas from computers to materials science. Important developments in Stealth aircraft, cruise missiles, small satellites, robotics, and artificial intelligence have also benefited from agency support (Broad 1991; Pollack 1989a). The Defense Advanced Research Projects Agency (DARPA)—its name was changed in 1972—has no labs of its own, but instead supports university and industry research.[4] According to one commentator, "the agency works by searching the land for hints of promising research and then paying select companies and universities to push back high-risk, high payoff frontiers" (Broad 1991: B5). The organization is viewed as bureaucratically lean, and much of its success is attributed to its organizational flexibility, the amount of discretion it is permitted, and its ability to provide resources for research with few strings attached (Edelson and Stern 1989: 8; Pollack 1989a).

As national economic competitiveness emerged as an important concern in policy debates in recent years, DARPA came into the limelight. Its success in promoting the development of critical technologies led to calls for reorienting the agency more toward the development of dual use technologies than it was in the past. In the early years of the Cold War, military R&D—particularly DARPA supported research—had civilian spinoffs, but this was more by accident than design (Pollack 1989a). More recently, and perhaps since the 1970s, there has been growing recognition in the agency and beyond it that national security depends upon a strong economy, and that the agency should support key technologies to keep U.S. industry competitive (Pollack 1989a; Marshall 1991: 22).

If NASA solved the problem of promoting civilian space research and DARPA aimed to improve the coordination and development of sophisticated defense-related technology, the most significant executive branch effort to improve federal research policymaking capacity in the wake of Sputnik was the establishment of the Office of Science and Technology (OST) within the Executive Office of the President. The office was created in June of 1962 when President Kennedy's Reorganization Plan Number Two became effective in the absence of Congressional disapproval (Pursell 1985: 28).

According to historian Jeffrey Stine, "OST was intended to coordinate Federal science policy—a role NSF had failed to achieve. . . . OST absorbed the science policymaking responsibility from NSF, although the Foundation was still charged with providing staff support and science policy proposals and recommendations to OST" (Task Force on Science Policy 1986a: 47, 48; see also Brooks 1986: 22). OST worked with two other executive branch bodies—the Federal Council for Science and Technology and the President's Science Advisory Committee—in its coordinating efforts. But according to one commentator who watched science policy closely in the 1960s, there is some doubt as to whether OST was ever particularly effective in carrying out its coordination mandate (Greenberg 1967: 18).

OST was abolished by Richard Nixon through his Reorganization Plan in July of 1973. In May of 1976, Gerald Ford signed into law an act creating the Office of Science and Technology Policy (OSTP) to replace OST. The OSTP was expected to have responsibility for advising the president and the executive on science and technology policy matters. The office was intended to assist the Office of Management and Budget (formerly the Bureau of the Budget) in reviews of proposed budgets for federal R&D programs and to promote stronger partnership between various actors in the scientific enterprise (Katz 1980: 229).

With the Federal Council for Science and Technology, later the Federal Coordinating Council for Science, Engineering, and Technology (FCCSET), OSTP has been the agency responsible for coordinating research policy. According to one analyst, however, this system has not been very effective in promoting coordination. Until its replacement in late 1993, FCST/FCCSET members represented various federal agencies, and since decision making was based on consensus, only vague recommendations were made, not coordinated national policy (Katz 1980: 234).[5] In addition, OSTP's coordination and planning capacity is severely limited because the agency has no budget control. Indeed, according to some sources, during the Bush administration the agency came into conflict with OMB, the organization that controls financial resources (Stiles interview, 1/29/92).

In keeping with the tradition of Harley Kilgore, the period immediately after the launching of Sputnik did not pass without an effort to *centralize and coordinate* research policy. Early in 1958, Senator Hubert Humphrey introduced S. 3126, entitled the Science and Technology Act of 1958. The act called for the establishment of a cabinet-level Department of Science and Technology. NSF, the Commerce Department's Patent Office, the Office of Technical Services, the National Bureau of Standards, the Atomic Energy Commission, and various programs of the Smithsonian were to be transferred to it. The act provided for the creation of several National Institutes of Science Research within the department. Each institute was to be responsible for a different field of basic research. The department secretary was given discretion with regard to making provision for applied research (Pursell 1985: 19).

Humphrey saw his proposal as providing much needed "coordination and centralization of the now dispersed federal [government] scientific activities" in the interest of national security (Humphrey 1960: 27). Like so many other efforts at *coordination and centralization,* this one failed. The effort was opposed by existing federal agencies interested in protecting their turf, scientists interested in maintaining a pluralist funding system, and people inside and outside government opposed to the expansion of government (House Task Force on Science Policy 1986: 45).

With the exception perhaps of NASA, understanding the capacities of and constraints upon these Cold War agencies is important for understanding the likely future of research policy in the United States. DARPA and OSTP have been at the center of recent debates over reorganizing research policy. DARPA certainly is an institutional legacy which supporters of greater coordination, planning, and industry-government collaboration in research policy have drawn on and on which they may continue to draw. As for OSTP, without the capacity to coordinate budgets across government departments, its planning capacity is severely limited and depends to a great degree upon the interests of the administration in power. According to Katz, OSTP was promoted by Ford, but fared less well under Carter. Katz contends that under Carter "The OSTP was blocked from a proactive role . . . largely because the President's top advisors recognized that many areas of science and technology were politically sensitive and hence should be handled at the political level" (1980: 230). The OSTP director during the Bush administration had a longstanding interest in greater coordination and proactive policy, but his capacity to act was clearly constrained by the others in the administration. President Clinton seems fully behind his OSTP chief, and the office was responsible for release of the administration's high profile technology policy statement (Clinton and Gore 1993).

Cold War rhetoric meant an orientation toward science for national prestige and security. The structure of the state continued to constrain federal capacity for policy coordination. Fragmentation of responsibilities restricted the capacity of the OSTP to play a significant coordination and planning role in research policy. And in the absence of historically institutionalized responsibilities, OSTP's policymaking role has waxed and waned with the interests and inclinations of succeeding administrations.

Proposals for Organizational Change in the 1980s

Two prominent congressional proposals to reconfigure the federal system for making research policy in the 1980s tell us a good deal about what has changed and what has not since Kilgore's original proposals in the early

1940s. In 1989, Democratic senator John Glenn introduced a bill that would have reorganized the Department of Commerce to give it central responsibility for the promotion of civilian technology. The act called for the creation of a National Advanced Civilian Advisory Board to advise the department secretary on the technical and economic merit of each request for support and on whether to require an industry commitment for matching funds. It called for representation on this board by business, academia, NSF, DARPA, and state and local governments. In assessing requests for support, the act required the department secretary to coordinate policy with other government agencies with interests in technology policy (U.S. Senate 1990: 127–38). The motive for the bill was the finding that there "is a critical need to link policies which enhance competitiveness, technology development, and national security" (92). More specifically, Senator Glenn argued that "one of the sources of this Nation's competitiveness problem is that the structural organization and functions of some of the Federal Government's key agencies are simply out of step with the dramatic changes which have occurred in the American economy in the last decade, and in the nations around the world" (1).

Within the department, Glenn's bill (S. 1978) called for the establishment of an Advanced Civilian Technology Agency (ACTA), a civilian DARPA (*Science* 1989). The act outlined two primary functions for ACTA. First, the agency would be responsible for promoting and assisting "in the development of advanced products, processes, and services" (U.S. Senate 1990: 125–26). In addition, the agency would support generic R&D by business, academic institutions, and private laboratories intended to "advance civilian technological developments and facilitate the more rapid commercialization of new products, processes, and services based on such developments" (126). Agency support for R&D projects was expected to come in the form of grants, contracts, and cooperative agreements. Finally, in an effort to mimic the reputed bases of DARPA's success, the legislation explicitly called for a small staff for ACTA (only forty persons) and required that employees be recruited primarily from industry and not make a long-term career out of their employment with the agency. The point of this provision seems to have been to maintain organizational flexibility and promote innovation.

Hearings held on the bill focused largely on the proposal for an Advanced Civilian Technology Agency. Testimony by former government officials, representatives of high-tech industry, and senators was generally positive. As one witness remarked, "I think the need for a civilian ARPA is crucial and urgent. I would liken the creation of a civilian ARPA in the face of the Japanese technology challenge of 1990 to the creation of NASA in the face of the Soviet space challenge in 1958" (U.S. Senate 1990: 29).

Witnesses did have some concerns. Some expressed worry that the agency might become a "pork barrel," and others urged that the organization be politically insulated. In an environment in which industrial policy had become associated with an unacceptable role for the government in "picking industrial winners and losers"—that is, with a significant government role in organizing the economy—several witnesses stressed the importance of convincing conservatives that the proposed agency would not be "an instrument of industrial policy" (U.S. Senate 1990: 66).

Despite these concerns and the administration's unwillingness to take a position on the bill, Glenn's staff was generally happy with the hearings. One aide noted that the hearings produced "very good testimony in support of the bill" (Weiss interview, 1/30/92). But what happened to the bill after the hearings says a good deal about the barriers to organizational reform for research policymaking in light of the fragmented character state and the lack of discipline and coordination in American political parties. The Glenn forces attempted to attach their proposal to the Omnibus Trade bill, which they believed would be passed early in the congressional session, but Glenn's efforts were hampered by Senate Commerce Committee Chair Ernest Hollings. One Glenn aide suggested that this opposition was the result of "turf problems" (Weiss interview, 1/30/92). Glenn was proposing reform of the agency for which Hollings' committee was responsible.

When the bill came to the Senate floor, Senator Hollings offered an amendment to strike the ACTA provisions from the bill on the grounds that ACTA would be too costly (Weiss interview, 1/30/92). Senators voted overwhelmingly to remove the ACTA provisions from the bill, despite the fact that several programs to improve commercial technology development remained. Glenn aide Leonard Weiss gives this assessment of the defeat:

> It's many times easier for somebody who doesn't know a great deal about a particular issue to vote in favor of striking something that costs money than

to vote for it. And especially when a powerful standing committee chairman, like Hollings, is making a pitch and a lot of people have legislation that goes through the Commerce Committee. He has a lot of political chits that he can call on, and so he won pretty handily. (interview, 1/30/92)

What Weiss is saying is truly revealing about the nature of policymaking in the United States. Despite broad agreement that the U.S. economy has serious structural problems, there was no consensus, even among Senate Democrats, about what to do. Weiss's point concerning the likelihood that other senators would need to bring bills before Hollings and did not want to offend him points to the free-for-all character of congressional policymaking. There was no plan Democrats could get behind prior to the trade bill reaching the Senate floor. Congressional politics is atomized, and what is more individual committee chairs have sufficient power to block legislative efforts on their own. Of course, even if Glenn's proposal had passed the Senate and the House, it would have confronted still another structural impediment to innovation: a Republican president opposed in principle to any significant government role in the economy. Even in cases where technology policy efforts have not suffered a presidential veto, in one important instance, President Bush avoided implementing important technology policy legislation for several years.

A second effort to reorganize research policymaking in the late 1980s says more about the benefits and limits of policy legacies than about the constraints imposed on policy innovation by a fragmented state. In 1987, House Science Committee Chair Democrat George Brown proposed two new science and technology policy agencies. Brown proposed creation of a federal Department of Science and Technology in HR 2164. The bill aimed to "advance the national prosperity and welfare" through the creation of the new department (U.S. House of Representatives 1987: 542). The department was to be constructed by taking pieces from NSF and the Commerce Department. It would be responsible for promoting technology transfer and technology development. It would have significant responsibilities in the area of data collection and analysis, providing assessments of the social impacts of new technologies, how policies and other factors affect technology development, and on the appropriate allocation of resources for technology development (552–54). In promoting technology development, the department would use grants, contracts, and cooperative agreements, and the department had a mandate to coordinate its policies and programs with other federal programs, as well as private sector and state and local governments (563).

Brown's new department was to include an Advanced Research Projects Administration, and the administration's policy would be established by a National Technology Board composed of twenty-four members, appointed by the president, who are "eminent in the fields of business, labor, research, new product development, engineering, law, education, management consulting, environment, international relations, and public affairs" (U.S. House of Representatives 1987: 554). Sometimes referred to as the Advanced Research Projects Foundation in legislation, this administration would aim to promote research "and other activities to lay the groundwork for the development and use by United States industry of advanced and innovative manufacturing and process technologies" (558). Research and other activities supported by the administration would be conducted jointly with consortia of U.S. industry and would aim to solve generic problems of specific industries with the objective of making those industries more competitive (558).

At the same time that Brown introduced this legislation, he also proposed establishing a National Policy and Technology Foundation (HR 2165). The aim of this foundation was to improve and facilitate policymaking in an effort to improve the quality of life throughout the country, national economic performance, and trade competitiveness. Like his department proposal, this foundation would have significant data collection and assessment functions. Data would be collected to permit assessment of long-term consequences of existing and proposed public and private policy (U.S. House of Representatives 1987: 289). An important aim of the foundation was to create *coherent, highly visible policy.* Also like his proposed department, the foundation would have input on policymaking from representatives of industry, labor, academia, and government. Independent councils would be created to provide advice, and a board representing a broad range of social interests would establish foundation policies, review foundation budgets and programs, approve or disapprove all large grants and contracts, and report to the president and the Congress (343–47).

Like the Science and Technology Department bill, this bill was a response in part to concern about the effects of organizational fragmentation and the incoherence of science and technology policy on the social and economic well-being of the country. To redress the problem of agency fragmentation, the bill called for the transfer of a wide range of government programs—from agencies including the National Science Foundation, the Commerce Department, and the Department of Defense—which deal with applied research and technology development, to the foundation. And the foundation was expected to formulate national *mid- and long-range policies* on national comparative advantage, industries of the future, basic technology and R&D to bolster U.S. comparative advantage, generic science and technology, and partnerships for economically relevant R&D (U.S. House of Representatives 1987: 323–30).

Unlike Senator Glenn, Representative Brown never expected his proposals to be enacted. According to William Stiles, legislative director of Brown's House Committee on Science, Space and Technology at the time, Brown "tends to be anywhere from five to ten years ahead of reality. I don't think he had any ideas that a Department of Science and Technology would succeed when he was proposing it" (interview, 1/29/92). Instead, by promoting his proposals and holding hearings Brown and his staff aimed to broaden congressional and public debate. As Stiles remarked: "Brown has always been willing to be a stalking horse on these things and sort of propose things that are . . . outside the realm of possibility in order to widen the debate" (interview, 1/29/92).

Still, Brown's efforts are interesting because they hark back to the legacy of Harley Kilgore. Like Kilgore's efforts, Brown's proposals call for creation of powerful central agencies to coordinate policy efforts. Like Kilgore's legislation, Brown's suggestions emphasize long-term planning and consideration of the social and economic effects of proposed technologies. Finally, perhaps even more than Kilgore, Brown's work emphasizes the need for social consensus on policy initiatives and calls for institutionalizing a voice for representatives from a broad range of social interests in technology policymaking.

Transforming Technology Policy from Bush to Clinton

In the days since the NSF was established, there have been many efforts to centralize research policymaking (Dupree 1963; Pursell 1985). After

Sputnik, Senator Hubert Humphrey proposed creation of a Department of Science and Technology, but instead of one agency several new agencies were created. Further fragmentation or alternatively institutional inertia seems as likely today as it was after World War II and after Sputnik. Institutionally, little has changed. Political parties do not adhere to programs and do not enforce party discipline. Individual senators and congresspeople can go it alone. And, as the case of the Glenn proposal illustrates, congressional politics is itself highly fractured, a free-for-all. As well, division between the executive and the legislature and within the executive make more than ad hoc program creation unlikely.

The years of the Bush presidency make the constraints imposed by a fractured atomized state abundantly clear. Two major initiatives are pointed to by legislative and business advocates of a new technology policy that would promote national economic competitiveness. The first is the Critical Technologies Institute (CTI). The institute was the brain child of Democratic senator Jeff Binghaman. CTI was created in the Defense Authorization Bill of 1990. The aim of the institute is in part to rationalize federal research support for critical technologies. To achieve this aim, a principal task of the CTI, which the legislation specified should be housed in the Office of Science and Technology Policy (OSTP), is to develop a list of technologies critical for U.S. economic competitiveness and national security. Every two years the list is supposed to be updated. In addition to developing this list, the legislation creating CTI gives the institute a role in aiding OSTP's development of a federal investment strategy for critical technologies (Hamilton 1990). Oversight of the institute is to be provided by a twenty-one member board of trustees chaired by the director of OSTP and including representatives from major federal science agencies (including NASA, NSF, the Department of Defense, the Commerce Department, the Energy Department, and the Department of Health and Human Services). In addition, the board is supposed to include ten representatives from industry and academia.

The Bush White House initially opposed CTI, and in July 1991, Bush Science Advisor Allan Bromley asked the Congress to take the $5 million appropriation for CTI back (*Science 1991:* 1343). According to Bromley, "it . . . became clear that what Congress had in mind was development of policy road maps for critical technologies. We felt these might be better produced by the private sector than at OSTP" (quoted in Lepkowski 1991:

4). Ultimately, the administration changed its tune on CTI, and President Bush's last budget provided a million dollars for this program. According to one White House official, the administration's view in its waning days was "let's try it out and see what it does for us" (Perrolle interview, 1/31/92).

Denise Michel, senior manager at the American Electronics Association (AEA), suggests that pressure from Senator Binghaman and the AEA led to the administration's change of heart. But there seems to have been at least one intervening factor: a shift in power on science policy issues in the administration. OSTP Deputy Director Perrolle suggests former Chief of Staff John Sununu had a background in science and played a useful role in science policymaking within the administration and that OMB chief Darman had a strong commitment to technology policy (interview 1/31/92). But Sununu, Darman, and economic advisor Michael Boskin are widely regarded as favoring a nongovernmental laissez-faire approach to technology policy (Branscomb 1991). Indeed, in April 1990 DARPA Director Craig Fields, a proponent of targeting the development of economically strategic technologies, was fired (Corcoran 1990: 82).

The power center on technology policy in the Bush administration shifted with the firing of Sununu, and Allan Bromley, never an advocate of a laissez-faire approach to research policy, appears to have gained a stronger position in the administration. This political drama suggests again the power of individuals in a fractured and atomistic state. A simple shift among those who the president pays attention to can determine the success or failure of a congressionally mandated program.

A second program looked at hopefully by proponents of a new technology policy, a policy that will promote the development of generic technologies, is the creation of the National Institute of Standards and Technology (NIST), particularly the Advanced Technology Program (ATP) at NIST. NIST was formerly the National Bureau of Standards (NBS), but, as a result of the Omnibus Trade and Competitiveness Act of 1988, the name of the NBS was changed, as was its focus. In addition to serving as the nation's laboratory for standards and measurement, NIST added to its mission responsibility for promoting the competitiveness of U.S. industry (U.S. House of Representatives 1988: 930). In the case of the Advanced Technology Program this means provision of grants and other forms of funding to individual firms and consortia of firms to support research "beyond the basic research stage but before commercialization of a new product" (Ember 1990: 17). The idea is to promote the development of technologies which might otherwise be underfunded (McClenahen 1990: 51). Often these are generic technologies or the stage of research is pre-competitive.[6] Since firms fear failing to recover their investment in such technologies, research in such areas is likely to be undersupported. As with CTI, the Bush administration was lukewarm to the Advanced Technology Program. According to Denise Michel of the American Electronics Association, the administration opposed the ATP even once it signed legislation creating it and refused to request funds for about two years. Michel suggests that some in the administration viewed ATP as industrial policy, a program intended to pick industrial "winners and losers" (interview, 1/31/92).

The trajectory of ATP mirrors that of the Critical Technologies Institute. For several years the Bush administration opposed the program, though they got behind it toward the end of Bush's term in office. Bush's final budget called for a thirty-six percent increase in the Program (Bromley 1992: 10). AEA's Michel attributes the change in policy to efforts by her organization and a coalition of business groups called the Advanced Technology Coalition. She contends that AEA and the Coalition "beat . . . them into submission. . . . I believe the administration reached the point where they felt it was a losing battle and would cost them more than it was worth to continue to oppose it. I think they've relaxed their concern about industrial policy a little bit" (interview 1/31/92).

Whether industry pressure is responsible for this change in policy is unclear. Certainly, however, the shift in power among administration proponents and opponents of technology policy that is more than R&D tax credits says a good deal about the fragmented and undisciplined character of the executive. Stability of actors' power bases in the executive is not institutionally protected, and the administration is not constrained to fulfill a party program—thus it is more easily swayed by powerful actors within the administration or pressure from interest groups from without. As these factors shift, so does policy.

The election of a new administration in November 1992 marked a move away from the largely laissez-faire policies of the Bush administration toward the kind of post–Cold War private-public cooperative technology policy advocated by many in high-tech industry. Indeed, according to Bloch and Cheney, the Clinton administration has brought technology policy to the "center of the national economic policy agenda" (1993: 55). The administration supports unprecedentedly high levels of funding for the Advanced Technology Program (see Anderson 1993: 1245; Cramer 1993; Mervis et al. 1993). In addition, with the end of the Cold War, the Clinton administration plans a central role for the Defense Advanced Research Projects Agency in defense reconversion and the development of dual-use technologies. In keeping with its expanded mission, the administration supported returning DARPA to its original name—Advanced Research Projects Agency (ARPA)—and the name change was given effect through the 1993 Defense Appropriations Act (Branscomb and Parker 1993: 89).[7]

During the 1992 presidential campaign, Bill Clinton proposed creation of a new agency—a civilian DARPA similar to the Glenn proposal—to foster development and commercialization of critical technologies (Chapman 1992: 47–48). But this initiative was nowhere to be found in the administration's first technology policy statement in February 1993 (Clinton and Gore 1993). According to Clinton's science advisor, Jack Gibbons, the plan now is to spread DARPA-like programs across the government (Anderson and Norman 1993: 1116).

In a discussion of future trajectories of U.S. technology policy, Branscomb suggests that "Americans are quite pragmatic about institutional innovation and tend to prefer adapting existing institutions to new policies rather than trying to restructure government" (1993b: 282). But this pragmatism cannot explain the transformation of Clinton's campaign proposal concerning a civilian DARPA; instead, *political* pragmatism seems superior as a partial explanation for Clinton's decision. As I illustrated in earlier chapters, given the structure of the state and society in the United States, constructing new—especially comprehensive—federal agencies is a difficult task. It is much easier to make incremental changes in existing agencies, and this ultimately is what the Clinton administration decided to do.

In my earlier discussion of the Bush administration, I suggested the importance of individuals in shaping policy, given the structure of the executive branch. Here again, Clinton's choices and the associated change in policy direction are consistent with the general themes of this book and are interesting in their own right. Clinton selected Laura Tyson, a technology expert, to head his Council of Economic Advisors. William Perry, a former electronics executive and Pentagon official, was tapped to be Deputy Secretary of Defense with responsibility for ARPA. Perry has long been an advocate of using the military budget to promote dual-use technologies, and he favored dropping the *D* from DARPA (Marshall 1993a: 1818). Finally, although he ultimately asked to have his nomination for defense secretary to replace Les Aspin withdrawn, Bobby R. Inman has publicly supported government promotion of dual use technologies (see Bingaman and Inman 1992; Inman and Burton 1991). Clinton ultimately selected Perry to become defense secretary.

Continuity and Change in Science and Technology Policy

If little has changed organizationally in terms of the structure of the state, political parties, and state-society relations, the current so-called crisis of competitiveness has altered the interests of some actors and consequently the alliances possible between actors, and perhaps ultimately the relative balance of power between actors. When World War II came to an end, elite scientists sought to guarantee government support for basic research. Vannevar Bush and his vanguard argued that basic research was the necessary foundation for a strong economy, and in light of the wartime success of the OSRD—and, of course, the Manhattan Project—scientists' claims were broadly supported. With their industrial allies and given their industrial background, Bush and his colleagues saw no need for government to support applied or technology research. Indeed, support for such research would be an inappropriate use of government resources, they believed, because it would put government into direct competition with industry. Representatives from firms with research capacity, however, agreed with scientists that basic research was an important foundation of national economic welfare, and they advocated government support for basic research. Such research was not profitable for firms to support since they could not prevent "free riders" from benefiting from the research they funded.

Today, many leaders of the science elite make similar arguments to those made by Vannevar Bush nearly a half century ago. In a report issued in January 1991, Leon Lederman, then-president of the American Association for the Advancement of Science (AAAS) and a Nobel prize–winning physicist, argued that academic research is in crisis. In the report, *Science: The End of the Frontier?,* Lederman relates the results of an informal survey of his academic colleagues. He found that their morale was universally low, and the source of discontent was inadequate funding for research that is now more expensive by degrees of magnitude than the smaller scale research from days gone by.

Through its title, the report clearly intends at once to contrast the heady days of the post–World War II period with the current climate for science and to remind us of the myriad of benefits science has brought this country. Lederman claims that "America has lived and grown great through science and technology" (1991: 4). He argues that "we have hitched our economy to the best scientific research system we could develop and have prospered as a result" (4). Unlike Vannevar Bush, who clearly argued that "basic" science is the bedrock of economic advance, Lederman seems careful not to claim too much credit for basic, as against "applied," research. But the former AAAS president's focus on "academic science" and his allusions to the importance of the "understanding of nature" and "new research" make clear who Lederman is speaking for (4, 17). His is a call for the government to increase support for basic research, and this cry is justified by the role that academic—read basic—research will play in improving the social and economic welfare of this country.

Like Vannevar Bush, Lederman seems to suggest that the ideal environment for this support is one in which scientists control the allocation of resources. Lederman asserts that the preferred situation is "one in which any talented scientist can obtain funding if he or she has a good idea and can meet the burden of reasonable review and resistance" (1991: 14). He contrasts this system with one in which government agency officials determine which areas of science to fund. Thus, by implication we must assume that Lederman intends scientist-peers to determine what counts as a good idea.

Lederman's view is echoed, though with more nuance, in a 1993 report issued by the National Academy of Sciences with the National Academy of Engineering and the Institute of Medicine. The report asserts that the context within which science and technology will develop is different from that when Bush's *Science—The Endless Frontier* (*SEF*) was written; but while the report's authoring committee acknowledges that science and technology are not sufficient to guarantee progress, they are necessary, and in keeping with an elite science discourse, this report suggests that the centrality of serendipity in fundamental research means that making such work goal-directed would ultimately be detrimental to national welfare (18). The report continues with this logic, suggesting that the United States should be among the world's leaders in "all major areas of science" because when researchers "are working at world levels in all disciplines, they can bring the best available knowledge to bear on problems related to national objectives, even if the knowledge appears unexpectedly in a field not traditionally linked to that objective" (19). In keeping with an elite science position, the report suggests that peer scientists should assess research performance. According to the committee authoring the report, "scientists immersed in a particular field are best qualified to appraise the true quality of the work being done, to identify the most promising and exciting advances, and to project the status of the field into the future" (22).

Unlike the Lederman report, written by an elite scientist and having the imprimatur of a scientist advocacy group, this report is not intended to reflect the views of the NAS alone—the nation's most elite group of scientists—but also the views of the nation's prominent engineers. In keeping with this fact, the second part of the document focuses on the development of technology. Here, the report recognizes the change in context since publication of *SEF.* While acknowledging a federal role in the development of commercial technology, it sees the private sector as having primary responsibility (National Academy of Science et al. 1993: 35).

A third interesting case of scientists trying to hold the line within a changing political environment comes from Harold Varmus, a Nobel prize winner and current National Institutes of Health chief, and a former president of the American Society for Cell Biology, Marc Kirschner. In a 1992 opinion piece in the *New York Times,* these two prominent scientists argued against moves toward priority setting and targeted research, suggesting that this approach is not the best way to find cures for devastating diseases. Instead, like Lederman and Vannevar Bush before him, Varmus and Kirschner argue that "the most effective long-term approach to improving health lies in fostering the research that increases understanding of genes and tissues" (A15).[8]

But the position advocated by these science elites is less convincing today than it was in the aftermath of World War II. Indeed, some scientist leaders recognize that their heyday may be over and have recommended retrenchment. While Lederman called for doubling the federal budget for science over the next several years, some segments of the scientific community have called, following the recommendation of National Academy of Sciences (NAS) President Frank Press, for priority setting by field in order to head off a public role in priority setting. According to astronomer John Bahcall, "astronomers have recognized that if they do not set their own priorities, then funding agencies and congressional officials will do it for them" (1991: 1412). Space scientists met in 1992 to determine priorities for their own field in the face of possible cuts in the NASA budget (Marshall 1992: 527).

As for business, it is clear to many that it is not greater support for basic research as such that is needed, but research in areas with clear economic importance, especially research on generic technologies. Recall the statement of Inman and Burton at the beginning of this chapter: "When it comes to technology, U.S. public policy can no longer afford to be preoccupied with basic research and military issues; economic security and industrial competitiveness are also vital considerations" (1991: A19).

Like the problem with industry support for basic research recognized by Vannevar Bush's allies at the time, firms are hesitant to support precompetitive research on their own since they will be unlikely to reap exclusive benefit from this research. In addition, the short-run orientation of American firms makes any long-term investment unwise and unprofitable.

It was industry—high technology especially—that was the principle social interest behind the Critical Technologies Institute and the Advanced Technology Program. The Advanced Technology Coalition headed by the American Electronics Institute, a leading trade group pushing for increasing government support for a comprehensive technology policy, has criticized the technology policymaking system in the United States as "fragmented, ad-hoc . . . [and this] has left the U.S. unable to focus resources most effectively and . . . establish long term national goals." In a letter to then presidential Chief of Staff John Sununu and other members of President Bush's staff, the coalition called on the administration to support "coordinated, long-term federal research initiatives" in critical technologies.[9]

While industry is not adverse to increased funding for basic research (indeed, the Advanced Technology Coalition, as well as the Council on Competitiveness, calls for increased support for NSF), their focus is clearly elsewhere. In a 1991 report, the industry-dominated Council on Competitiveness explicitly called for a break from the philosophy underlying Vannevar Bush's 1945 *Science—The Endless Frontier Report* (1991: 16).[10] The council's report advised the government to develop a five-year plan to increase support for critical generic technologies. They recommended a move toward increased cooperation between the executive and industry, a dual-use focus for DARPA, a reorientation of federal laboratories toward national technology needs, more technology policy coordination, and improved procurement policies. As well, the council called on the government to help create an environment that would promote investment and innovation in technology. This would involve reviewing policies that affect business time horizons, promoting policies to reduce the cost of capital for the development of priority technology, and removing barriers to industrywide cooperation (1991: 45–48).

With the election of Bill Clinton, the discourse of basic science that dominated most discussions of science and research policy during the postwar period was clearly displaced by the technology policy discourse articulated by high-tech industry in the United States. In a report representing the Clinton administration's first major technology policy statement, the president, like the Council on Competitiveness, called for an explicit break with the philosophy underlying postwar science and technology policy:

> American technology must move in a new direction to build economic strength and spur economic growth. The traditional federal role in technology development has been limited to support of basic science and mission-oriented research in the Defense Department, NASA, and other agencies. This strategy was appropriate for a previous generation but not for today's profound challenges. We cannot rely on the serendipitous application of defense technology to the private sector. We must aim directly at these new challenges and focus our efforts on the new opportunities before us, recog-

> nizing that government can play a key role in helping private firms develop and profit from innovations. (Clinton and Gore 1993: 1)

The report follows a logic articulated by representatives of high-tech industry during the Bush administration and at the same time is consistent with Kilgore's efforts to promote coordination and planning in federal science and technology policy. Clinton calls for "the coordinated management of technology across all the government . . . [,] a closer working partnership among industry[,] . . . governments, workers, and universities," and focus of national efforts on crucial technologies (1).

With high-tech industry, the administration does not have an interest in short-changing basic science research. Indeed, in a phrase that undoubtedly warmed the hearts of the likes of Leon Lederman and Harold Varmus, the report calls for a reaffirmation of "our commitment to basic science, the foundation on which all technical progress is ultimately built" (Clinton and Gore 1993: 1; see also p. 24). At the same time, this report flatly rejects the idea that economic advance will result from a technology policy that depends exclusively on support for unfettered basic research. The report does not view science as the "engine" of economic growth as Vannevar Bush did; it is technology that will provide the basis for an improved standard of living for Americans (7). Clinton's report calls on the nation to "harness technology so that it improves the quality of [American's] . . . lives and the economic strength of our nation" (2).[11]

Leon Lederman clearly recognized the difficulty the United States has had in converting research into something of economic importance (1991: 18–19). And business clearly recognizes the importance of basic research. But the interests of industry and academic scientists do not mesh as well as they did after World War II. What this divergence of interest will mean for the future of industry-scientist alliances and the political power of each group is not certain. But there are important differences between the present crisis of competitiveness and World War II that suggest a reconfiguration of science-industry relations may already be occurring. Scientists could lay claim to solving or resolving the wartime crisis. Scientists today cannot be blamed for the current crisis, but there is no evidence, either, that the rhetoric of Lederman and others is convincing to the public or policymakers. Industry, on the other hand, certainly, in part, can be blamed for the current crisis, but, as well, the few bright spots on America's economic horizon are the result of high-tech advances: computers and biotechnology particularly. And policymakers from the White House to the Capitol clearly view the future in high-technology terms. Thus, while in the creation of the NSF, scientists—or at least a science vanguard—were clearly the senior partners in a science-business alliance, with the impact of industry largely being funneled through them, today the academic science community may find itself a junior partner in a new alliance. Increasingly, a rhetoric demanding support for basic research in the interest of national economic welfare may not be enough.

Conclusion

From World War II to Sputnik to the current economic crisis of competitiveness, it is clear that historical junctures widely viewed as crises motivate organizational change. It is equally clear that the structure of the state, state-society relations, and political parties place limits on the possibilities for organizational change. The structural character of the American state and society have changed very little since World War II, and all evidence suggests that the creation of a single powerful new agency—or even dramatically increased policy coordination and planning—is unlikely in the near future. From World War II through Sputnik to the present, crisis has prompted incremental, ad hoc, and fragmented organizational changes. Single agencies are created for limited purposes, and the organization of the executive makes policymaking unstable and limits planning capacity.

In contrast to earlier periods, one thing does appear to be changing in the arena of research policy, and that is who the important actors are. After World War II, Vannevar Bush and his vanguard could do little to shape the organizational form of the National Science Foundation—to create a powerful central agency. They were constrained by the range of organizational factors that I have repeatedly pointed to. But with high symbolic capital, the permeable character of the state, and their ability to get a place within it, Bush and his colleagues were able to shape the philosophy underlying the system of postwar research policymaking. With their business allies, Bush and his colleagues pushed for support of basic research and control by scientists. If early technology policy moves by the Clinton administration are any indication, the successful policy initiatives for the duration of this "crisis of competitiveness" are likely to have a content shaped to a significant degree by high-tech industry. This will mean an increasing orientation toward promoting government business alliances in the development of generic technologies—as against basic research—and perhaps an institutionalized role for business interests in defining the priorities of a new research policy. But it will not mean more centralization, serious coordination, and planning.

NOTES

1 Thinking about the Politics of Science and Science Policy

1. In keeping with the theoretical undercurrent of this work, I understand *science* to involve all the activities required for the production of empirically based "knowledge." These activities include *research*—investigation through experimentation, observation, and so on—as well as a range of other activities, including priority setting and funding activities, peer evaluation, and training.
2. It is questionable whether it is possible to draw a clear line between the purely technical and the purely social. From one perspective, technology is social to its very core. At the very least, what is to count as a "technological" shortcoming must be seen as a product of social processes. I put the word "technological" in quotation marks here because I wish to point to the problem of calling a phenomenon strictly technological. At the same time, I wish to bracket the entire problem of defining technology and the line between the technological and the social. For some recent approaches to the sociology of technology, see Bijker et al. 1989 and Latour 1987.
3. Lederman's report has come in for some serious criticisms. These criticisms are outlined and responded to by Kleppner 1991.
4. The competitive position of the United States in critical technologies appears to have improved in the past several years (Council on Competitiveness 1994).
5. Letter from J. Richard Iverson, president of the American Electronics Association, and leaders of six other trade groups to John Sununu, Chief of Staff to President George Bush, 11/21/91. Available from the American Electronics Association.

6. Often critics of the system refer to it as fragmented, and proponents call it pluralistic. I use the term *fragmented* because it is consistent with recent social science literature on the state and state capacities. See, among others, Evans et al. 1985b and Hall 1986.
7. The claim that "science is a field like any other" constitutes only half of Bourdieu's argument; he argues further that all the characteristics shared with other fields—a distribution of power, struggles, interests, and so forth—take a particular form in science. In addition, although it is less clearly articulated, Bourdieu seems to suggest that, unlike other fields, science has the potential to be ruled by reason. See Bourdieu 1975, 1991.
8. For other work in the science studies area on science policy see Kwa 1987 and MacKenzie and Spinardi 1988.
9. My discussion of Cambrosio and his colleagues in this chapter draws on my published response to their study. For an extended critique of the paper by Cambrosio et al. 1991, see Kleinman 1991. My disagreements with Cambrosio and his colleagues have led to a debate on how best to analyze science policy. See Abraham 1994, Cambrosio et al. 1991, and Wynne 1992.
10. To respond in part to Cambrosio, Limoges, and Pronovost's (1991) reaction to my position and to anticipate criticisms from proponents of the actor-network approach more generally, my claims about the limitations of this approach do not amount to a return to what they might see as a reductionistic "classical" sociology. I do not intend to suggest that structures, or their characteristics and effects, can be assumed a priori. These characteristics and effects are matters for historical investigation, which must be addressed for each particular case. Nevertheless, structures have histories of their own and are fixed at any given point in time. While we should certainly not make assumptions about the effects of structures a priori, we must inevitably begin investigation with the historically established characteristics of structures, which we assume from the outset to be of importance. We may be wrong as to their importance, but that can only be determined once analysis is under way.
11. For reviews of the neo-Marxist literature on the state, see Carnoy 1984, Gold et al. 1975, and Jessup 1982.
12. I believe that understanding the specific characteristics of the American case is facilitated by understanding the differences between the American case and the cases of other countries. And, while a formal comparative analysis is beyond the scope of this study, my research is "comparatively informed" (Evans et al. 1985b: 349). My recognition of the importance of specific institutional dimensions of the American state and society is drawn from a wide range of comparative work on social and economic policy. As well, in chapter six I undertake a comparative reconnaissance to bolster the counterfactual case I pose throughout this study.

13. Although the aim of this study is not to illustrate the role of the American state in facilitating capital accumulation, it is clear that the postwar research policy debates traced here do address crucial questions about capital accumulation; in fact, the stakes concern the manner in which the state will facilitate capital accumulation. In the debates leading to the establishment of the NSF in 1950, industry recognized this and sought legislation that would work to their advantage. Representatives from business were clear that they did not want the state competing with them in undertaking applied research; they did see a role for government in promoting basic research, an area that was generally not considered a cost-effective investment for individual firms. In addition, business representatives strongly supported granting property rights for inventions from state-supported research to business. Finally, as I show in chapter seven, current research policy debates focus explicitly on the role of the government in supporting research to enhance the profitability of U.S. industry.

14. I do not dispute the claim by neo-Marxists that the United States is a capitalist country in which, to an important degree, social relationships are structured by one's position in the relations of production. I would argue, however, that contemporary U.S. society is characterized by multiple centers of power: power is decentered, not defined only in terms of one's position in production relations. During and after World War II, elite scientists' power was based on a mix of social and cultural capital, the transformation of one into the other, and the transformation of a combination of the two into powerful organizational positions for the group's leaders in the state. For a more detailed discussion of this issue, see chapter three.

15. Evans and his colleagues mention the importance of informal networks. See Evans et al. 1985b: 356.

16. Early in setting the agenda for state-centered research, Skocpol admitted the importance of class, but this admission did not lead her to pay serious attention to the independent role of class in explaining policy outcomes (1985: 20). Recently, however, Skocpol and her colleagues have adopted a more balanced approach to the study of politics. Terming their approach the "institutional political process" approach, these analysts now argue that "political struggle and policy outcomes are . . . jointly conditioned by the institutional arrangements of the state and by class and other social relationships, but never once and for all" (Weir, et al. 1988: 16–17).

17. For a brief discussion of the rise and fall of the corporate elite research tradition, see Quadagno 1991.

18. This central observation has been recognized recently, but without much self-consciousness by analysts in the state-centered and more broadly institutionalist traditions.

19. For a detailed discussion of the concept of scientific field, see chapter two.
20. I draw my inspiration for the concept of credibility from Bourdieu's idea of capital (1984). In thinking about several types of capital—social, cultural, and economic—Bourdieu uses an economic metaphor: he talks about the investment of different forms of capital to produce a profit. Thus, one might use one's social connections—social capital—to win a contract for one's company. Or one might invest one's cultural capital—in this case, academic credentials—to obtain a job. In both cases, we can think of these investments yielding a profit. In addition, we see that in realizing the profit, the form of capital has been transformed from social and cultural to economic (i.e., money).

The initial credibility (status or reputation) elite scientists had in their relationships with political leaders in the early 1940s was based on social connections and on academic and professional credentials (social and cultural capital). Viewed schematically, they were able to use this credibility (invest it) to secure positions in the government during the war, and these positions allowed them to accumulate further credibility. In positions of government power, elite scientists like Vannevar Bush had access to extensive government resources, which were invested in the development of war technologies. The success of these technologies improved the status and reputations of the working scientists, but by extension it also increased the credibility of elite scientific administrators. Credibility accumulated during the war could be invested: transformed into social capital (meetings with members of the government) or exchanged for political favors of various types.
21. Although the battle to establish the National Science Foundation in an immediate and direct way involved only a handful of elite scientists, the sense I wish to create is of a larger group. The group of elite scientists—led by Vannevar Bush—active in the NSF policy debate represented the interests of the scientists who received most of the money from the federal government during the war and were likely to continue to be the beneficiaries of the federal largesse if Bush had his way after the war. These scientists came from a limited number of universities and were concentrated in the physical sciences.
22. As I remark in note 1, the notion that anything is strictly technical is problematic; the development and reception of any technology occurs in a social context. In some very basic sense, then, technology is fundamentally social. When I refer to technical achievements in this context, however, I am referring simply to technological artifacts: radar, penicillin, and so on.
23. Some historians contend that *the* high point of elite scientists involvement in postwar science policymaking came with the elevation in status accorded the President's Science Advisory Committee (1957) under President Eisenhower (Geiger 1993: 95).

2 Mapping Science: The Scientific Field in the United States, 1850–1940

1. Early work in the sociology of the "scientific community" treats science as a distinct and autonomous social sphere. This perspective pays little attention to the resource requirements of scientific research or to the institutional context within which scientific work is embedded (see Hagstrom 1965 and Merton 1973). Even work critical of the Mertonian tradition often fails to consider the institutional context in which research is embedded, and this is especially true of the laboratory studies of the late 1970s and 1980s (see Latour and Woolgar 1979). Philosophers and scientists sometimes make a normative plea for the autonomy of science (see Bush 1960 [1945], 1970, Lederman 1991, and Polanyi 1951, 1962).
2. In what follows, although I talk about a progressive change in the organization of research in general, I do not intend to leave the impression that all scientific disciplines and specialties had identical experiences. While the historical record does indicate a general trend toward organizational transformation with increased resource dependency, the experiences of fields of inquiry varied across a range of dimensions. Whitley 1984 provides an interesting discussion of this variation. See also Fuchs 1993.
3. I draw my inspiration for the concept of scientific field from Pierre Bourdieu (1975, 1991), but my usage of the term should not be taken as equivalent to his. Although Bourdieu's conceptualization has room for a view of the field as a shifting configuration constituted by the overlap and interaction of several institutional spheres, his emphasis is largely on power/authority relations between scientists, and in this sense is closer to Merton's (1973) scientific community than my notion of scientific field.
4. For an interesting discussion of mimetic processes in organizational development, see DiMaggio and Powell 1983.
5. Beyond the resources necessary for the development of the modern U.S. research university, Carol Gruber outlines several material-social conditions that precipitated the restructuring of higher education in the United States. Gruber suggests that "the acceleration of urbanization and industrialization and the settlement of the continent created a demand for scientific and technical knowledge by both business and the federal and state governments" (1975: 12). In addition, according to Gruber, original and experimental work in science and engineering was gaining respect, and the products of that research served as a challenge to the classical curriculum, the position of which was weakened "by an erosion of religious influence and an advancing secularism" (12). Finally, not only were resources provided by the great capitalists and their philanthropies, but also growing government recognition of the importance of higher education made public funds available to universities.

6. Universities are typically viewed as the bastions of basic or fundamental research (Dupree 1957: 297; U.S. House Task Force on Science Policy 1986a: 9), but these are slippery concepts and certainly change with the motivation of the person deploying them. Roger Geiger, a historian of the American research university, refers to the commitment of early twentieth-century research universities to "the advancement of knowledge for its own sake" (1986: 159). In practice, however, one could think of the development of a commercial product as motivated by a desire for "knowledge for its own sake." Certainly, although firms may convince their scientists to do certain work with the claim that they are pursuing knowledge for its own sake, the logic of the market requires firms to be concerned with profit and thus typically with material products. Thus, if the ultimate aim of research is producing commodities, can the research then be called "basic"?

There are, of course, other ways of distinguishing types of research. *Basic,* or fundamental, research is often seen as the building block upon which more *applied*—product oriented—research is based. It is understood to explore the fundamental "laws of nature." Again, this distinction is problematic since often the line between such fundamental and applied research is very thin indeed. Recent developments in such industries as biotechnology illustrate this quite clearly.

Instead of asking about the motivation of the researcher, we can ask about the requirements of the research funder. Does the funding entity ask for anything specific? Sometimes analysts refer to directed and nondirected research. *Directed* work is typically seen to be akin to applied research and *nondirected* work to basic research. But, again, direction versus nondirection may tell us whether knowledge is being pursued for "its own sake," but it tells us little about the relationship of the research to production (i.e., whether it is applied).

Even if it were possible to make these distinctions more clearly, participants and analysts may not use the same criteria in labeling. So what is "basic" in one context may be "applied" in another. Labeling research is thus inevitably fuzzy. Consequently, in this chapter and throughout this study, I typically use a combination of the labels used by participants in the history I explore and labels that can be derived from ascertaining the motivation behind the research.

7. According to Gruber, the "ideal of service," which had a formative influence on the development of American universities, is reflected in the establishment of the land grant system (1975: 29–30).

8. Significantly, according to Kloppenburg, there was widespread opposition to the idea of land grant universities among farmers. They feared, Kloppenburg argues, that "education would teach them nothing about farming they did not already know and yet cost them much" (1988: 58). The impetus for the practical training orientation of the land grants "found its origins and firmest commitment

in the industrial sector, which saw an opportunity to use the sale of public land to finance the training of a skilled manufacturing work force" (59).

9. This portrayal should not be exaggerated and may vary by period, industry, and firm size. Such a portrait seems to me likely to characterize a period before high-technology industry and more likely to be found in large companies. Where firms need scientists to undertake fundamental or foundational research quite far from the commodity form—as is the case in many new high-tech industries—management may provide scientists with an environment that approximates the academic research environment as closely as possible. Scientists may be encouraged to publish research and present academic papers, thus keeping them in touch with the newest ideas. They may, in addition, be given a certain amount of time each week to work on a research topic of their choosing, a topic with no immediate relevance to the firm. See Dubinskas 1985 and Whalley 1986.

10. Birr provides a slightly different set of numbers; he says there were approximately 1,000 industrial research laboratories in 1927 and 1,769 in 1938 (1966: 69).

11. It is not possible to determine the precise real increase in the number of research personnel between 1920 and 1940, but part of the increase in the number of research personnel reported by the National Research Council may be an artifact of improved firm reporting and expansion of the category of research personnel. The actual increase in research personnel employed during the period may have doubled instead of having increased fivefold (National Research Council 1940: 174–76).

12. While this statement that scientists had majority control of NACA is contradicted by Dupree, who notes that majority control was in the hands of government officials, he concludes that "civilian scientists were in practice predominant" (1957: 334).

13. Although many scientists involved in the war effort during World War I were inducted into the armed forces, scientists involved in administration of NACA were civilians.

3 A Scientists' War: Institutional Advantage, Social Connections, and Credibility

1. There is widespread recognition that crises (e.g., wars and economic depressions) create opportunities for institutional change (Hall 1986; Ikenberry 1988; Krasner 1984; Skowronek 1982). McLauchlan specifically contends that world wars are transformative events (1989: 83), and Tilly argues that the most impor-

tant factor in shaping states is war and preparation for war (1990). Tilly also suggests that an increase in the scale of government is associated with war (1975). More specifically, Hooks suggests that World War II precipitated the transformation of the U.S. state from a social security orientation to a national security orientation (1991: 11, 50). McLauchlan adds that in the wake of World War II national security states are science-intensive (1989: 87).

2. Since publication of Stephen Skowronek's work, the definition of state building has come to mean more than the addition of new state agencies. Skowronek views state building as the "systematic transformation of an entire mode of governmental operations" (1982: 14–15) and traces the transformation of the U.S. state in the late nineteenth century from a state dominated by courts and parties to one with expanded national administrative capacities. More recently, employing Skowronek's definition of state building, Gregory Hooks (1991) explores the transformation of the U.S. state from its domestic orientation during the New Deal to a national security orientation during and after World War II. The transformation that resulted from creation of the National Science Foundation does not constitute state building in the comprehensive sense that the term is used by Skowronek, Hooks, and other scholars. At the same time, Senator Harley Kilgore's initial proposal for a postwar science and technology agency promised a radical transformation and expansion of the federal role in science and technology policy. What is more, even though Kilgore was not successful, the struggles over science and technology policymaking institutions waged between World War II and 1950 fundamentally shaped the contours of the federal role in the science and technology area in the postwar period.

3. One need not take a position on the character of knowledge to subscribe to this contention. That is, whether knowledge reflects something "real" or is merely a resource—a kind of capital not directly reflecting the "real"—is inconsequential for the basic argument.

4. By *elite scientific research,* I mean research associated with elite institutions. In using the term, I do not assume that elite institutions are the best in any objective sense; however, they are certainly the institutions that dominate science.

Many scientists would probably claim that the institutions I call elite—Harvard, MIT, Johns Hopkins, Caltech, and others—have earned the designation based on their record of producing the "best science." Bush and others would certainly have argued that on universalistic (i.e., merit-based) criteria, the best research was done at elite universities. He and others would have claimed that the quality of research done at these institutions is what explains why they were the institutions that received the most government support for research during the war. Indeed, these scientists did argue that the government should support the "best science." Supporting research on this basis, they believed, would have the unfortunate result that elite institutions would continue to get a disproportionate amount of federal support.

There is no reason to believe that peer review—evaluation of one's work by scholar colleagues—leads to promotion of the "best science." Decisions may be based on nonmerit criteria—intellectual trends, who one knows, and so on—which are set by scientists in control of the resources. Thus, scientists who dominate fields may be in a position to define the elite and contribute to their reproduction in ways that do not mesh with the ideology of universalism which pervades science. See Bourdieu 1975; Crane 1965; Kevles 1987.

5. Vannevar Bush Papers, general correspondence, box 110, file 2617: Teeter, John (Sept. 1947–May 1949), Bush to Teeter, 6/3/49; box 57, file 1381: Johns Hopkins University (1941–46), Bush to Mrs. John W. Garret, 12/3/46; box 70, file 1718: MIT (1947–49), abstract of minutes of meeting of corporation, 12/1; box 96, file 2210: Research Corporation (1941–44); box 15, file 352: Brookings Institution, Moulton to Bush, 5/4/44; box 83, file 1896: NRC—Committee on Policies, minutes of meeting, 4/22/40, and Jewett to Committee on Policies, 5/24/44.

6. Bush Papers, general correspondence, box 15, file 352: Brookings Institution, Moulton to Bush, 5/4/44.

7. Vannevar Bush Papers, general correspondence, box 51, file 1269: Hopkins, Harry, "Organization of Defense Research," memo with notation: "As sent to Harry Hopkins," 3/31/41.

8. FDR Papers, president's secretary's file, box 2, safe: V. Bush.

9. FDR Papers, president's safe file, box 2, safe: V. Bush, "Report of the NRDC," 6/28/41.

10. FDR Papers, official file, box 1, file 4482: OSRD, Blanford to FDR, 6/27/41.

11. The concept of *scientific community* is problematic. It sometimes is used to speak of a homogeneous "community" of scientists, and distinctions by disciplines and distributions of authority and power are often overlooked. In addition, this concept can reinforce the idea of science as an autonomous social space. I use the concept in a less encompassing way to refer to the population of scientists in the United States. In doing so, I do not wish to suggest all disciplines or scientists are the same.

12. FDR Papers, president's secretary's file, box 121, subject file: Bush, V., FDR to Oppenheimer, 6/29/43.

13. Bush Papers, general correspondence, box 55, file 1375: Jewett, Frank B. (1941), Bush to Jewett, 4/4/41; FDR Papers, official file, box 1, file 4482: OSRD, FDR to Malmberg, 12/11/44; FDR Papers, president's secretary's file (subject), box 121, file Vannevar Bush: FDR to Bush, 6/23/42, FDR to Bush, 12/28/42, FDR to Bush, 2/2/43, Bush to FDR, 8/28/44, FDR to Bush, 9/29/44, Bush to FDR, 3/7/44.

14. According to Kevles, Admiral Julius Furer, who was a naval member of the OSRD advisory council, "suspected Bush of primary interest in the advancement of Vannevar Bush" (1987: 313).

15. Bush Papers, general correspondence.

16. FDR Papers, president's secretary's file (subject), box 121, file: Vannevar Bush, Stimson to FDR, 10/18/44.

17. That these views were widespread among opinion leaders is made clear by a cursory look at congressional hearings, as well as contemporary news accounts. It should be noted, however, that Jones's (1975, 1976) general thesis is that these attitudes were widespread among the American public at large. However, since Jones's study is based only on analysis of politicians' speeches and newspaper articles, his inference that the attitudes he documents were widespread among the American public at large is unwarranted. While this might be the case, he provides no evidence to support his assertions.

18. Not all leaders had such a positive image of scientists and their work after the war. Eleanor Roosevelt, for example, was suspicious of the selfish motives of some OSRD-affiliated scientists. See FDR Papers, president's secretary's file (subject), box 121, file: Vannevar Bush, FDR to Bush, 3/9/45.

4 High Hopes: Setting the Agenda in the Battle for a Postwar Research Policy

1. On the science intensive national security state, see McLauchlan (1989).

2. This is a major goal of many of the technology policy proposals tendered in the 1980s and early 1990s.

3. In a sense, calling for an agency with the capacity to *selectively* promote technological development makes Kilgore's proposal a precursor to recent calls for an industrial policy in the United States.

4. See chapter five for a more detailed discussion of the AASCW and its role in the debate over postwar research policy.

5. Bush Papers, general correspondence, box 56, file 1376: Jewett, Frank B. (1942), Senator Wallace White to Jewett, 12/14/42.

6. Historian Carroll Pursell argued that Jewett missed the point that during the period under discussion American life was already being "rationalized," and he correctly argues that the real question was whether the blueprint for the reorganization of science "should be drawn up by politicians answerable to the people or by private enterprises answerable only to themselves" (1979b: 374).

7. Bush Papers, general correspondence, box 56, file 1376: Jewett, Frank B. (1942), Jewett to Bush, 11/16/46.

8. Bush Papers, general correspondence, box 56, file 1376: Jewett, Frank B. (1943), Jewett to Bush, 5/27/43.
9. See also Kilgore Papers, A & M 967, series 4, box 1, folder 2.
10. Since World War II, scientists and science-based corporations have often used a discourse that stresses the distinctiveness of science as a social sphere in order to protect their autonomy. They have argued that the meritocratic and self-governing character of science entitles scientists to autonomy and, indeed, that autonomy is necessary for the optimal functioning of science. Given the supposedly special character of science, scientists and science-based corporations have stressed the inappropriateness of public involvement in decisions that affect science. For a discussion of a contemporary episode which illustrates these issues, see Kleinman and Kloppenburg 1991.
11. Kilgore Papers, A & M 967, series 4, box 1: Kilgore to Arthur Halsted, 11/2/43, and Kilgore to Victor Lebow, 12/23/43.
12. Minutes, NAM Board of Directors meeting, reel 3, 5/4/43, "Appendix D—S. 702, and an Analysis of it by the National Association of Manufacturers," p. 1, 4.
13. NAM Papers, accession 1411, ser. 5, committee minutes, box 9, 1944 (Oct.)–1945 (July), File: NAM committee minutes, 1945, April, "Report of Meeting of NAM Committee on Patents," 4/11/45.
14. NAM Papers, minutes, NAM Board of Directors meeting, 5/4/43, reel 3.
15. There has been a great deal of controversy over the origins of the Roosevelt letter. One early historical account suggests the letter came directly from President Roosevelt (Penick et al. 1972 [1965]: 15). As Bush tells the story himself, he was talking to the president one day, and Roosevelt asked him what would happen to science after the war. Bush responded that the prospects didn't seem good, and the president asked him to write a report. Bush claims to have then drafted a letter which the president signed (Bush 1970: 10). More recent historical accounts tell the story somewhat differently. Lomask (1973) argues that presidential assistant and Bush aide Oscar Cox came up with the idea for the letter, and the letter embodies Cox's concerns. Kevles believes Lomask is generally on the right track, but says the letter also reflected the concerns of Bush and many of his OSRD colleagues (1974: 800). Finally, the minor role Roosevelt played in pushing the letter seems to be supported by the diaries of his budget chief, Harold Smith. After a conference with the president, Smith concluded that "apparently he [Roosevelt] had quite forgotten having signed any communication asking Bush to make a study." Although it is not certain that the study referred to is *Science—The Endless Frontier,* the date of the diary entry (3/23/45), in between the date of the letter requesting the report and the release of the report, makes this likely. Harold Smith Papers, box 3, conference with President Roosevelt, 1943–45, March 23, 1945.
16. FDR Papers, official file, box 1, file 4482: OSRD, Bush to Rosenman, 2/27/45.

17. FDR Papers, official file, box 1, file 4482: OSRD, Rosenman to Johnson, 1/1/45, and Bush to Rosenman, 12/20/44.
18. Papers of the Directors of Industrial Research, Hagley Museum and Library, accession 1851, box 2, secretary's files, 1942–46, general correspondence, 10/22/42–6/27/46: attachment to letter, O. E. Buckley to Major, 1/8/45. H. Alexander Smith Papers, box 132, file: Labor and Welfare Committee, NSF, Smith to Merck, 2/11/47, and Smith to Saltonstall, 2/10/47; Smith to Cole, 3/13/47; Smith to Buckley, 3/13/47.
19. NAM Papers, Board of Directors meeting, 5/4/43, appendix D, microfilm, reel 3.
20. Shapley and Roy suggest that Bush favored "central planning for science" (1985: 43). The evidence presented in this book and elsewhere does not support this contention.
21. It is worth noting that in its discussion of what kind of research should be supported, Bush's *SEF* throws a couple of bones to Kilgore and his supporters. The report asserts a need for research on housing, an area of concern to Kilgore. In addition, the report asserts the need to develop the weaker research institutions throughout the country, a concern expressed throughout the debate by Kilgore's allies (Bush [1945] 1960: 12, 16, 96).
22. The Bowman committee went further than Bush himself, recommending that the NRF "assist industry and business, particularly small enterprises, in establishing research facilities and in obtaining scientific and technical information and guidance, in order to expedite the transition from scientific discovery to technological application" (Bush [1945] 1960: 117).
23. Kilgore's and Bush's proposals for the organization of government support for scientific research in the postwar period were at the center of debate, but other proposals were made. Frank Jewett, a close friend of Bush's, as well as a colleague at OSRD and president of the National Academy of Sciences and Bell Labs, opposed both Kilgore's and Bush's plans, fearing that government involvement in science would lead to its politicization (Penick et al. 1972: 133; Bush Papers, general correspondence, box 55, file 1375: Jewett, Frank B. (1941), Jewett to Bush 3/5/41). Instead, Jewett favored private support for scientific research which he claimed could be facilitated by changes in the tax laws (Chalkley 1951: 25). Jewett was behind several pieces of legislation that embodied his philosophy of the appropriate relationship between science and government (Rowan 1985: 113–14), but while Jewett did have some supporters among elite scientists, he never had substantial support (76).
24. FDR Papers, official file, box 1, folder 4482: OSRD, 1941–45, FDR to Carl Malmberg, 12/11/44.
25. See FDR personal file, Kilgore (8970).

5 Toward Peace on the Potomac: State Building and the Genesis of the National Science Foundation

1. See note 23 in chapter four for a brief discussion of the proposals made by Frank Jewett.

2. The issue of geographic distribution of resources for science versus distribution based on "best science" criteria has roots dating back at least to federal science funding debates that occurred in the years after World War I (Kevles 1987: 151).

3. In the U.S. state, policymaking is decentralized; the opposite of a decentralized state is one in which policymaking is centralized. In the latter, single-executive agencies are responsible for given policy areas, and lines of policy jurisdiction are clearly delineated.

4. The opposite of a permeable state is an insulated state—a state protected from ad hoc interest-group influence.

5. For a discussion of the concept of state strength, see Evans et al. (1985b). See also Katzenstein 1978, Krasner 1978, and Nettl 1968.

6. States are not always uniformly strong or weak. A state may have great policymaking and implementation capacity in one area and not in another. Hooks (1991) argues, for example, that while many components of the U.S. state are weak, the Pentagon emerged from World War II with something akin to a sectorial-level industrial policymaking capacity.

7. See Kilgore Papers, A & M 967, series 8, box 1, folder 6.

8. NAM Papers, accession 1411, series 5, committee minutes, Committee on Patents and Research, 4/28/48.

9. NAM Papers, accession 1411, series 1, box 205, file: Government—House of Representatives, "Statement by Ira Mosher, Chairman, National Association of Manufacturers before a Sub-committee of the Judiciary Committee of the House of Representatives, May 6, 1946."

10. See NAM Papers, Board of Directors minutes, microfilm.

11. See NAM Papers, accession 1411, series 5, box 10: committee minutes, 1946, March–April; NAM Committee on Patents and Research minutes, March 6, 1946.

12. "Report of Meeting of N.A.M. Committee on Patents," 4/11/45, Biltmore Hotel, New York, NY; Papers of NAM, accession 1411, series 5, committee minutes, box 9, 1944 (Oct.)–1945 (July), File: NAM committee minutes.

13. "Minutes of Meeting of NAM Committee on Patent and Research," 6/26/46, Waldorf Astoria Hotel, New York, NY; NAM Papers, accession 1411, series 5, committee minutes, box 10: 1945 (Aug.)–1946 (Dec.); File: NAM committee minutes, 1946, May–Sept. 30.

In the end, NAM believed that it did have influence on the property rights provision of the 1949 NSF legislation. However, I have found no evidence to confirm the organization's influence. See NAM Papers, "Progress Report of the NAM Committee on Patents and Research," by H. C. Ramsey, appendix E to NAM Board minutes, 12/6/49, reel 5.

14. NAM Papers, accession 1411, series 5, committee minutes, box 10, file: NAM committee minutes, 1946, March–April, minutes of executive group of NAM Committee on Patents and Research, 4/24/46; business report for period 12/2–12/31/47, microfilm; accession 1411, series 2, "Official Program, NAM—52nd Annual Congress of American Industry," December 3–5, 1947, box 322, file 47-3057.

15. NAM Papers, accession 1411, series 5, committee minutes, box 6, minutes of meeting of NAM Committee on Patents and Research, 9/24/42.

16. Papers of the Directors of Industrial Research, accession 1851, secretary's files, 1923–42, box 1: "Directors of Industrial Research: Brief History."

17. Papers of the Directors of Industrial Research, accession 1851, box 2, secretary's files, 1942–46, general correspondence: 10/22/42–6/27/46, Major to Howard, 4/7/45.

18. Papers of the Directors of Industrial Research, accession 1851, box 2, secretary's files, 1946–52, general correspondence: Weith to Austin, 11/19/46.

19. Papers of the Directors of Industrial Research, accession 1851, box 2, secretary's files, 1942–46, general correspondence: 10/22/42–6/27/46, Major to Mees, 9/19/44, and Jewett to DIR, 3/11/46.

20. For a discussion of world wars as transformative events, see McLauchlan 1989.

21. Twenty-five may be an underestimate of the overlap in membership between DIR and NAM. I was unable to obtain official membership lists for NAM, and have used a combination of lists including board and committee membership and annual dinner participation. See NAM Papers, Hagley Museum and Library, accession 1411.

22. Hooks 1991 shows that the World War II mobilization led to the demise of the social security state and the rise of the national security state. Kilgore's early proposals were clearly more compatible with a New Deal social security state. Furthermore, it is within the context of this transformation that both Kilgore and Bush dropped proposals for a military division in NSF under civilian control. Moreover, in this developing science-intensive national security state (McLauchlan 1989: 87) the resources for civilian agency-supported research would be dwarfed by military-supported research.

23. See Kilgore Papers, A&M 1108, box 2, file 5.

24. Bush Papers, general correspondence, box 29, file 654: Oscar Cox, 1945–49, Bush to Cox, 11/27/46.
25. See Bush Papers, general correspondence, box 110, file 2617: Teeter, John H. (1944–July 1947). See also H. Alexander Smith papers, box 132, file: Labor and Welfare Committee, NSF.
26. Bush Papers, general correspondence, box 85, file 1912: NSF (Jan.–Feb. 1947), Bush to Taft, 2/20/47.
27. Papers of Harold Smith, daily record 1945, box 3.
28. Kevles notes that the role given to Kilgore's proposed foundation to make recommendations concerning changes in government research activities did not make the agency the clear administrative control center for research policy since the recommendations were not binding, but advisory only (1977: 15).
29. NAM Papers, minutes of NAM Committee on Patents Executive Group, 11/20/45, accession 1411, series 5, committee minutes, box 10, file: NAM committee minutes, 1945 (Oct.–Dec.).
30. NAM Papers, accession 1411, series 5, committee minutes, box 10, file: NAM committee minutes, 1945 (Aug.)–1946 (Oct.), minutes of NAM Committee on Patents Executive Group, 9/18/45.
31. Kilgore Papers, University of West Virginia, A & M 967, box 1, folder 7: Kilgore to Gray, 3/14/46.
32. In addition to his concern to placate scientists, Kevles suggests that Kilgore was particularly interested in a passable bill, since he was up for reelection in the fall of 1946 (1987: 357). This may have strengthened his willingness to compromise.
33. Bush Papers, general correspondence, box 56, file 1377: Jewett, Frank B. (1947–49), Jewett to Bush, 3/29/46.
34. NAM Papers, NAM business report, 3/20/46–4/26/46, appended to NAM board minutes, 4/26/46, microfilm; accession 1411, series 5, committee minutes, box 10, file: NAM committee minutes, 1946, March–April, minutes of executive group, NAM Committee on Patents and Research, 3/5/46.
35. Bush Papers, general correspondence, box 110, file 2617: Teeter, John (1944–July 1947), Teeter to Bush, 5/17/46 and 6/10/46.
36. Bush Papers, general correspondence, box 21, file: Carnegie Endowment of Washington, presidential correspondence, Scherer to Bush, 6/17/47).
37. NAM Papers, NAM business report, 2/20–3/20/46, microfilm, reel 4; accession 1411, series 5, committee minutes, box 10, file: NAM committee minutes, 1946, March–April, meeting of the executive group of NAM Committee on Patents and Research, 4/24/46, and file: NAM committee minutes, 1946, May–Sept., minutes of NAM Committee on Patents and Research, 6/26/46.

38. H. Alexander Smith Papers, AM 19137, box 132, file: Labor and Welfare Committee, NSF, Bush to Smith, 12/23/46, and Smith to McCoy, 2/21/47.
39. H. Alexander Smith Papers, box 132, file: Labor and Welfare Committee, NSF, Kilgore to Smith, 1/30/47.
40. Bush Papers, general correspondence, box 85, file 1912: NSF (Dec 1946, H. A. Smith to Bush, 12/18/46, H. A. Smith to Bush, 1/6/47, Bush to H. A. Smith, 1/28/47; box 69, file 1686: Major, Randolph, Bush to Major, 12/30/46; H. Alexander Smith Papers, box 132 (Labor Committee, 1947–48), file: Labor and Welfare Committee, NSF, Smith to Taft, 1/6/47, Smith to Bush, 3/4/47, Smith to Conant, 5/29/47, Smith to Compton, 5/29/47.
41. Bush Papers, general correspondence, box 85, file 1912: NSF (Jan.–Feb. 1947), Bush to Conant, 1/29/47.
42. H. Alexander Smith Papers, box 132 (Labor Committee, 1947–48), file: Labor and Welfare Committee, NSF, Smith to Teeter, 2/11/47; Bush Papers, general correspondence, box 85; file 1912: NSF (Jan.–Feb. 1947), Bush to Conant, 1/29/47; file 1912: NSF (Dec. 1946, Teeter to Bush, 1/16/47).
43. H. Alexander Smith Papers, box 132, file: Labor and Welfare Committee, NSF, Smith to Forrestal, 2/11/47, 3/13/47, 3/28/47.
44. In understanding the dense quality of connections in the debate over the NSF, several things should be noted about this meeting. First, all but one of the companies represented at the dinner were members of the National Association of Manufacturers and several were represented on the NAM Committee on Patents at one time or another. Second, Bush and George Merck had a long history of contact: they were contemporaries as students at MIT and Harvard, respectively. Then, during World War II, Merck was a special consultant to the Secretary of War at the same time Bush headed OSRD (Bush 1970: 208–9). John Connor, special assistant to OSRD during the war, went to work for Merck as secretary and general counsel after the war. Bush himself later sat on the Merck board. For more on George Merck, see Young 1980: 447–48.

Bush Papers, general correspondence, box 69, file 1686: Major, Randolph, Major to Bush, 12/27/46; box 75, file 1751: Merck, George (1945–50), Major to Bush, 1/9/47, Bush to Major, 1/14/47; box 69, file 1686: Major, Randolph, Major to Bush 12/30/46, Major to Bush 1/6/47; H. Alexander Smith Papers, box 132, file: Labor and Welfare Committee, NSF, Smith to Major, 2/11/47, Major to Smith, 2/5/47.
45. Bush Papers, box 69, file 1686: Major, Randolph, Bush to Major, 12/30/46.
46. H. Alexander Smith papers, box 132, file: Labor and Welfare Committee, NSF, Smith to Merck, 2/11/47 and Smith to Saltonstall, 2/10/47.
47. H. Alexander Smith Papers, box 132, file: Labor and Welfare Committee, NSF, Smith to Cole, 3/13/47.

48. H. Alexander Smith Papers, box 132, file: Labor and Welfare Committee, NSF, Smith to Buckley, 3/13/47.
49. Kilgore Papers, box 48, file: NSF (79th and 80th Congresses), Inter-Society Committee meeting minutes, 2/23/47.
50. Bush Papers, general correspondence, box 110, file 2617: Teeter, John, Teeter to Bush, 11/12/47; box 85, file 1912: NSF (March–Dec. 1947), Bush to Webb, 12/27/47; H. Alexander Smith Papers, box 132 (Labor Committee, 1947–48), file: National Science Bill, Herman to Smith, 12/23/48; file: Labor and Welfare Committee, NSF, "confidential," 11/24/47, Teeter; Smith to Wolverton, 12/5/47; file: National Science Bill, "revised Draft-Section 6," 3/3/48; memo, Teeter to Smith, 3/24/48; Webb to Smith, 3/15/48; Shapley to Smith, 12/10/47.
51. H. Alexander Smith Papers, box 132 (Labor Committee, 1947–48), file: National Science Bill, memorandum, H. Alexander Smith Jr. to H. Alexander Smith, 3/17/48.
52. Kilgore Papers, box 48, file: NSF (79th, 80th Congress), memorandum, Science Legislation Study Group, Washington Association of Scientists, 4/5/48.
53. NAM Papers, accession 1411, series 5, committee minutes, box 11, 1947 (Jan.)–1948 (May), file: NAM committee minutes, 1948, Jan.–March, minutes of Exec Committee of NAM Committee on Patents and Research, 2/19/48; minutes of NAM Committee on Patents and Research, 2/20/48; minutes of NAM Subcommittee on the Government's Role in Research, 3/30/48; file: NAM committee minutes, 1948, July–Sept., minutes of NAM Committee on Patents and Research, 4/28/48; minutes of meeting of Subcommittee on the Government's Role in Research, 10/4/48.
54. NAM Papers, accession 1411, series 5, committee minutes, box 12, 1948 (June)–1949 (Oct.), file: NAM committee minutes, 1949, Jan.–Mar.
55. Bush Papers, general correspondence, box 110, file 2617: Teeter, John (Sept. 1947–May 1949), Teeter to Bush, 3/17/49.
56. Bush Papers, general correspondence, box 21, file: Carnegie Endowment of Washington, Scherer to Bush, 6/24/49.
57. Schooler does recognize that scientists are political actors who "seek influence on public policy making" (1971: 5).

6 From Grand Vision to Puny Partner: Fragmentation and the U.S. Research Policy Mosaic

1. McLauchlan (1989) explores the intimate linkages between federally funded science and national security concerns in the period after World War II.
2. Twelve years after the creation of the NSF, faced with the challenge of Sputnik,

President Kennedy clearly recognized that NSF was not equipped to coordinate science policies transcending agency lines. He created an Office of Science and Technology (OST) inside the Office of the President (Sherwood 1968: 611).

7 Possibilities and Prospects: Research Policy at a New Institutional Divide

1. Bromley was science advisor and head of the Office of Science and Technology Policy in the Bush administration.
2. For a discussion of the science-intensive national security state, see McLauchlan 1989.
3. According to Edelson and Stern, "although Sputnik was the proximate cause of DARPA's creation, it is important to note that there were important forces moving in the direction of a central military R&D unit. These included the military's steadily increasing need for high technology during and after World War II, and the evolution of central R&D bodies within the corporate structure of large companies. Sputnik, then, was responsible for the timing of and urgency attached to DARPA's creation, but the general need for such an institution can be traced to larger historical factors" (1989: 4).
4. With its name change, DARPA was established as a separate defense agency under the Office of the Secretary of Defense, and it came under the direction of a civilian (Edelson and Stern 1989: 6).
5. By executive order, President Clinton replaced FCCSET with a National Science and Technology Council (Holden 1993: 1643).
6. In a 1991 report, the Council on Competitiveness defines generic technology as "a concept, component, or process or the further investigation of scientific phenomena, that has the potential to be applied to a broad range of products or processes." Such a technology "may require subsequent research and development for commercial application." Precompetitive research and development are "activities up to the stage where technical uncertainties are sufficiently reduced to permit preliminary assessment of commercial potential and prior to development of application-specific commercial prototypes." At this stage in the research process "results can be shared within a consortium that can include potential competitors without reducing the incentives for individual firms to develop and market commercial products and processes based upon the results" (Council on Competitiveness 1991: 17).
7. There is a voluminous literature on developments in the Clinton administration's technology policy and alternatives to it. Good sources include *Science, Issues in Science and Technology,* and *Technology Review.* An especially thorough discussion of current technology policy issues can be found in Branscomb 1993. Eval-

uation of the effectiveness of various types of technology policies is beyond the scope of this chapter. Again, Branscomb 1993 is a useful source. See also Cohen et al. 1991, Lambright and Rahm 1992, and Shapley and Roy 1985.

8. For a more detailed version of this statement, see Bishop et al. 1993.

9. Letter from J. Richard Iverson, president of the American Electronics Association, and leaders of six other trade groups to John Sununu, Chief of Staff to President George Bush, 11/12/91. Available from the American Electronics Association.

10. While explicit calls for policies that break with the philosophy derived from *Science—The Endless Frontier* (*SEF*) have grown common in recent years, a 1985 book by Shapley and Roy suggests that the policy changes called for would not constitute a fundamental break with the Bush report philosophy, but only from the way the report came to be read in the years after its release. Most interpretations of the report—including this study—have viewed *SEF* as a call for basic research support: the basis for "the progress of industrial development" (Bush [1945] 1960: 18). This rendering of the report, shared by postwar policymakers and historians alike, ultimately led to policies that assumed new technologies would flow inevitably from basic scientific research.

Shapley and Roy suggest the Bush report "does not treat basic research in a vacuum, but as one of the steps in a chain of endeavor" (1985: 7). Although Bush and his colleagues may have recognized the importance of self-consciously linking basic scientific research to technological development, there is no textual evidence in *SEF* to indicate such recognition. Basic research is repeatedly referred to as the source of all scientific and technical progress in *Science—The Endless Frontier.* The report has very few references to applied research and contains no explicit discussion of the mechanisms by which basic research is transformed into industrial products and processes.

11. Calls for policies that establish links between basic research and technological development are commonplace today. See Branscomb 1993, and Shapley and Roy 1985, among others.

Locations of Archival Collections

Vannevar Bush Papers, Manuscript Collection, Library of Congress, Washington, D.C.; FDR Papers, Franklin Roosevelt Presidential Library, Hyde Park, NY; Kilgore Papers, University of West Virginia, Morgantown, VA; NAM Papers, Hagley Museum and Library, Wilmington, DE; H. Alexander Smith Papers, Princeton University, Princeton, NJ.

BIBLIOGRAPHY

Abbott, Andrew. 1988. *The System of Professions: An Essay on the Division of Expert Labor* (Chicago: University of Chicago Press).

Abraham, John. 1994. "Interests, Presuppositions and the Science Policy Construction Debate," *Social Studies of Science* 24: 123–32.

Abramson, Rudy. 1971. "Patron on the Potomac: the National Science Foundation," *Challenge* (May–June): 38–43.

Adams, Walter. 1972. "The Military-Industrial Complex and the New Industrial State." In Carroll Pursell (ed.), *The Military-Industrial Complex* (New York: Harper & Row), 81–94.

Alpert, Harry. 1955. "The Social Science and the NSF: 1945–55," *American Sociological Review* 12 (December): 653–67.

Amenta, Edwin and Theda Skocpol. 1988. "Redefining the New Deal: World War II and the Development of Social Provision in the United States." In Margaret Weir, Ann Orloff, and Theda Skocpol (eds.), *The Politics of Social Policy in the United States* (Princeton, NJ: Princeton University Press), 81–122.

American Association for the Advancement of Science. 1943. "Resolution of the Council on the Science Mobilization Bill (S. 702)," *Science* 98: 135–37.

American Institute of Physics. 1943. "The Mobilization of Science," *Science* 97: 482–83.

Amsterdamska, Olka. 1990. "Surely You Are Joking, Monsieur Latour!" *Science, Technology, and Human Values* 15(4): 495–504.

Anderson, Christopher. 1993. "Clinton's Technology Policy Emerges," *Science* 259 (February 26): 1244–45.

Anderson, Christopher and Colin Norman. 1993. "Jack Gibbons: Plugging into the Power Structure," *Science* 259 (February 19): 1115–16.

Atkinson, Michael M. and William D. Coleman. 1988. "Strong States and Weak States: Sectorial Policy Networks in Advanced Capitalist Economies." Unpublished paper, Department of Political Science, McMaster University, Hamilton, Ontario, Canada.

Auerbach, Lewis E. 1965. "Scientists in the New Deal: A Pre-war Episode in the Relations between Science and Government in the United States," *Minerva* 3 (summer): 457–82.

Axt, Richard G. 1952. *The Federal Government and Financing Higher Education* (New York: Columbia University Press).

Bahcall, John N. 1992. "Prioritizing Scientific Initiatives," *Science* 251 (March 22): 1412–13.

Barnes, Barry. 1974. *Scientific Knowledge and Sociological Theory* (London: Routledge & Kegan Paul).

Barrows, Clyde W. 1990. *Universities and the Capitalist State: Corporate Liberalism and the Reconstruction of Higher Education, 1894–1928* (Madison: University of Wisconsin Press).

Bartholomew, James R. 1989. *The Formation of Science in Japan: Building a Research Tradition* (New Haven, CT: Yale University Press).

Baumol, William J. 1989. "Is There a U.S. Productivity Crisis?" *Science* 243 (February 3): 611–15.

Baxter, James P. 1946. *Scientists against Time* (Cambridge, MA: MIT Press).

Bijker, Wiebe E., Thomas P. Hughes, and Trevor Pinch. 1989. *The Social Construction of Technological Systems* (Cambridge, MA: MIT Press).

Bingaman, Jeff and Bobby R. Inman. 1992. "Broadening Horizons for Defense R&D," *Issues in Science and Technology* (fall): 80–85.

Birr, Kendall. 1979. "Industrial Research Laboratories." In Nathan Reingold (ed.), *The Sciences in the American Context: New Perspectives* (Washington, D.C.: Smithsonian Institution Press), 193–207.

Birr, Kendall. 1966. "Science in American Industry." In David Van Tassel and Michael Hall (eds.), *Science and Society in the United States* (Homewood, IL: Dorsey Press), 35–80.

Bishop, J. Michael, Marc Kirschner, and Harold Varmus. 1993. "Science and the New Administration," *Science* 259 (January 22): 444–45.

Bloch, Erich and David Cheney. 1993. "Technology Policy Comes of Age," *Issues in Science and Technology* (summer): 55–60.

Block, Fred. "The Ruling Class Does Not Rule: Notes on the Marxist Theory of the State." In Fred Block, *Revising State Theory: Essays in Politics and Postindustrialism* (Philadelphia: Temple University Press), 51–68.

Bloor, David. 1976. *Knowledge and Social Imagery* (London: Routledge & Kegan Paul).

Blume, Stuart S. 1974. *Toward a Political Sociology of Science* (New York: Free Press).

Bourdieu, Pierre. 1991. "The Peculiar History of Scientific Reason," *Sociological Forum* 6(1): 3–26.

Bourdieu, Pierre. 1988. *Homo Academicus* (Stanford, CA: Stanford University Press).

Bourdieu, Pierre. 1984. *Distinction: A Social Critique of the Judgement of Taste* (Cambridge, MA: Harvard University Press).

Bourdieu, Pierre. 1975. "The Specificity of the Scientific Field and the Social conditions for the Progress of Reason," *Social Science Information* 14(5): 19–47.

Boyer, Paul. 1989. "'Some Sort of Peace': President Truman, the American People, and the Atomic Bomb." In Michael J. Lacey (ed.), *The Truman Presidency* (Cambridge, England: Cambridge University Press), 174–202.

Branscomb, Lewis M. 1993b. "Empowering Technology Policy." In Lewis M. Branscomb (ed.), *Empowering Technology: Implementing a U.S. Strategy* (Cambridge, MA: MIT Press), 266–94.

Branscomb, Lewis M. 1993c. "The National Technology Policy Debate." In Lewis M. Branscomb (ed.), *Empowering Technology: Implementing a U.S. Strategy* (Cambridge, MA: MIT Press), 1–35.

Branscomb, Lewis M. 1991. "Toward a U.S. Technology Policy," *Issues in Science and Technology*" (summer): 50–55.

Branscomb, Lewis M. and George Parker. 1993. "Funding Civilian and Dual-Use Industrial Technology." In Lewis M. Branscomb (ed.), *Empowering Technology: Implementing a U.S. Strategy* (Cambridge, MA: MIT Press), 64–102.

Brinkman, W. F. 1990. "A National Engineering and Technology Agency," *Science* 247 (February 23): 901.

Broad, William J. 1991. "Pentagon Wizards of Technology Eye Wider Civilian Role," *New York Times*, October 22, pp. B5, B9.

Broadhead, Robert S. and Ray C. Rist. 1976. "Gatekeepers and the Social Control of Social Research," *Social Problems* 23(3): 325–36.

Bromley, Allan. 1992. "Research and Development in the President's FY 1993 Budget." Statement of the Assistant to the President for Science and Technology, January 29. Copy available through the Office of Science and Technology Policy.

Bronk, Detlev W. 1975. "The National Science Foundation: Origins, Hopes, and Aspirations," *Science* 188: 409–14.

Bronk, Detlev W. 1974. "Science Advice in the White House," *Science* 186 (October 11): 116–21.

Brooks, Harvey. 1986. "An Analysis of Proposals for a Department of Science," *Technology in Society* 8: 19–31.

Brooks, Harvey. 1979. "The Problem of Research Priorities." In Gerald Holton and Robert Morison (eds.), *Limits of Scientific Inquiry* (New York: Norton), 171–90.

Brown, George E. 1988. "A New Institution for Science and Technology Policy-Making." In William T. Golden (ed.), *Science and Technology Advice to the President, Congress, and Judiciary* (New York: Pergamon), 65–70.

Brown, Harold and John Wilson. 1992–93. "A New Mechanism to Fund R&D," *Issues in Science and Technology* (winter): 36–41.

Bruce, Robert V. 1987. *The Launching of Modern American Science* (New York: Knopf).

Bulletin of the Atomic Scientists. 1950a. "FAS Statement on Science Foundation Bill," *Bulletin of the Atomic Scientists* 6 (June): 190.

Bulletin of Atomic Scientists. 1950b. "The National Science Foundation Act of 1950 [text of the Act]," *Bulletin of Atomic Scientists* 6 (June): 186–90.

Bulletin of Atomic Scientists. 1949. "Federation of American Scientists Announces Policy Decisions," *Bulletin of Atomic Scientists* 5 (June/July): 184–85.

Burch, Phillip H. Jr. 1973. "The NAM as an Interest Group," *Politics and Society* 4(1): 97–130.

Burton, Daniel F. Jr. 1992. "A New Model for U.S. Innovation," *Issues in Science and Technology* (summer): 52–59.

Bush, George. 1992. "Text of Bush's Message: Heating Up the Economy, and Looking Beyond," *New York Times,* January 29, p. A16.

Bush, Vannevar. 1970. *Pieces of the Action* (New York: Morrow).

Bush, Vannevar. [1945] 1960. *Science—The Endless Frontier: A Report to the President on a Program for Postwar Scientific Research* (Washington, D.C.: National Science Foundation).

Bush, Vannevar. 1943. "The Kilgore Bill," *Science* 98: 571–77.

Business Week. 1982. "A Technology Lag that May Stifle Growth," *Business Week* (October 11): 126–30.

Business Week. 1945. "Scientists Gain in Congressional Fight," *Business Week* (October 20): 7.

Cambrosio, Alberto, Camile Limoges, and Denyse Pronovost. 1991. "Analyzing Science Policy-making: Political Ontology or Ethnography? A Reply to Kleinman," *Social Studies of Science* 21: 775–81.

Cambrosio, Alberto, Camile Limoges, and Denyse Pronovost. 1990. "Representing Biotechnology: An Ethnography of Quebec Science Policy," *Social Studies of Science* 20: 195–227.

Cambrosio, Alberto, Camile Limoges, and Denyse Pronovost. 1991. "Analyzing Science Policy-Making: Political Ontology or Ethnography?: A Reply to Kleinman," *Social Studies of Science* 21: 775–82.

Campbell, John L. 1988. *Collapse of an Industry: Nuclear Power and the Contradictions of U.S. Policy* (Ithaca, NY: Cornell University Press).

Carey, William D. 1986. "An Idea Whose Time Has Not Come," *Technology in Society* 8: 77–82.

Carnoy, Martin. 1984. *The State and Political Theory* (Princeton, NJ: Princeton University Press).

Cawson, Alan. 1986. *Corporatism and Political Theory* (London: Basil Blackwell).

Chalkley, Lyman. 1951. "Prologue to the U.S. National Science Foundation (1942–1951)." Unpublished manuscript available in the Kilgore Papers, A & M 967, series 8, box 1, folder 1, University of West Virginia, Morgantown, West Virginia.

Chapman, Gary. 1992. "Push Comes to Shove on Technology Policy," *Technology Review* (November/December): 43–49.

Chemical Engineering. 1947. "Who Sank S. 526?" *Chemical Engineering* 54 (September): 134.

Chemical Engineering News. 1947. "Inter-Society Committee for a NSF," *Chemical and Engineering News* 25 (April 7): 972.

Cheney, David W. and William W. Grimes. 1991. *Japanese Technology Policy: What's the Secret?* (Washington, D.C.: Council on Competitiveness).

Clinton, William J. and Albert Gore. 1993. *Technology for America's Economic Growth, A New Direction to Build Economic Strength* (Washington, D.C.: U.S. Government Printing Office).

Coben, Stanley. 1979. "American Foundations as Patrons of Science: The Commitment to Individual Research." In Nathan Reingold (ed.), *The Sciences in the American Context: New Perspectives* (Washington, D.C.: Smithsonian Institution Press), 229–47.

Cohen, Linda R., Roger G. Noll, Jeffrey S. Banks, Susan A. Edelman, and William M. Pegram. 1991. *The Technology Pork Barrel* (Washington, D.C.: Brookings Institution).

Cohen, Stephen S. and John Zysman. 1988. "Manufacturing Innovation and American Industrial Competitiveness," *Science* 239 (March 4): 1110–15.

Compton, Karl T. 1943. "Organization of American Scientists for the War," *Science* 98 (2534): 71–76.

Congressional Quarterly. 1982. *Guide to Congress*, 3rd ed. (Washington, D.C.: Congressional Quarterly).

Corcoran, Elizabeth. 1993. "Computing's Controversial Plan," *Science* 260 (April 2): 20–22.

Corcoran, Elizabeth. 1990. "Talking Policy: The Administration Devises an Industrial Policy—Sort of," *Scientific American* (June): 82, 84.

Corson, Dale R. 1988. "The United States Has No Adequate Mechanism to Set

Long-Range." In William T. Golden (ed.), *Science and Technology Advice to the President, Congress, and Judiciary* (New York: Pergamon), 95–103.

Council on Competitiveness. 1994. *Critical Technologies: Update 1994* (Washington, D.C.: Council on Competitiveness).

Council on Competitiveness. 1991. *Gaining New Ground: Technology Priorities for America's Future* (Washington, D.C.: Council on Competitiveness).

Cozzens, Susan E. and Thomas F. Gieryn (eds.). 1990. *Theories of Science in Society* (Bloomington: University of Indiana Press).

Cramer, Jerome. 1993. "NIST: Measuring Up to a New Task," *Science* 259 (March 26): 1818–19.

Crane, Diana. 1965. "Scientists at Major and Minor Universities: A Study of Productivity and Recognition," *American Sociological Review* 30(5): 699–714.

Crawford, Mark. 1988. "Applied R & D Key for U.S. Trade," *Science* 241 (September 16): 1425.

Crease, Robert P. 1991. "Pork: Washington's Growth Industry," *Science* 254 (November 1): 640–43.

Culliton, Barbara J. 1989a. "NIH: The Good Old Days," *Science* 244 (June 23): 1437.

Culliton, Barbara J. 1989b. "Science Advisor Gets First Formal Look," *Science* 245: (July 21): 247–48.

Current Biography. 1940. "Bush, Vannevar," *Current Biography* 1(9): 13–14.

Curti, Merle and Roderick Nash. 1965. *Philanthropy in the Shaping of American Higher Education* (New Brunswick, NJ: Rutgers University Press).

Davis, Lance E. and Daniel J. Kevles. 1974. "The National Research Fund: A Case Study in the Industrial Support of Academic Science," *Minerva* 12 (April): 207–20.

Dickson, David. 1988. "Setting Research Goals Not Enough, Says OECD," *Science* 241 (August 19): 898.

Dickson, David. 1984. *The New Politics of Science* (New York: Pantheon).

DiMaggio, Paul J. and Walter W. Powell. 1983. "The Iron Cage Revisited: Institutional Isomorphism and Collective Rationality in Organizational Fields," *American Sociological Review* 48 (April): 147–60.

Domhoff, G. William. 1990. *The Power Elite and the State* (New York: Aldine De Gruyter).

Domhoff, G. William. 1983. *Who Rules America Now? A View for the '80s* (New York: Simon & Schuster).

Domhoff, G. William. 1979. *The Powers that Be: Process of Ruling Class Domination in America* (New York: Vintage).

Domhoff, G. William. 1974. *The Bohemian Grove and Other Retreats: A Study in Ruling-Class Cohesiveness* (New York: Harper & Row).

Dorfer, Ingemar N.H. 1975. "Science and Technology Policy in Sweden." In

T. Dixon Long and Christopher Wright (eds.), *Science Policies of Industrial Nations: Case Studies of the United States, Soviet Union, United Kingdom, France, Japan, and Sweden* (New York: Praeger), 169–90.

Dowd, Maureen. 1990. "Bush Appoints 13 to Science Panel," *New York Times,* February 3, p. 12.

Dubinskas, Frank. 1985. "The Culture Chasm: Scientists and Managers in Genetic Engineering Firms," *Technology Review* (May/June): 24–30 and 74.

Dubridge, Lee A. 1977. "Twenty-five Years of the National Science Foundation," *Proceedings of the American Philosophical Society* 121 (June 15): 191–94.

Dupree, A. Hunter. 1972. "The Great Insaturation of 1940: The Organization of Scientific Research for War." In Gerald Holton (ed.), *The Twentieth Century Sciences* (New York: Norton), 443–67.

Dupree, A. Hunter. 1965. "The Structure of the Government-University Partnership after World War II," *Bulletin of the History of Medicine* 39: 245–51.

Dupree, A. Hunter. 1963. "Central Scientific Organization in the United States Government," *Minerva* 1 (summer): 453–69.

Dupree, A. Hunter. 1957. *Science in the Federal Government* (Cambridge, MA: Belknap).

Edelson, Burton I. and Robert L. Stern. 1989. *The Operations of DARPA and Its Utility as a Model for a Civilian ARPA* (Washington, D.C.: Johns Hopkins Foreign Policy Institute). Reprinted in U.S. Senate, 1990, *Hearings before the Committee on Governmental Affairs on S. 1978,* 101st Congress, Second Session (Washington, D.C.: U.S. Government Printing Office).

Ember, Lois. 1990. "Commerce to Fund Advanced Technologies," *Chemical and Engineering News* (August 27): 17.

Encyclopedia Britannica, The New. 1991. "Radar," v.26, 15th Edition (Chicago: Encyclopedia Britannica).

England, J. Merton. 1982. *A Patron for Pure Science* (Washington, D.C.: National Science Foundation).

England, J. Merton. 1976. "Dr. Bush Writes a Report: 'Science—The Endless Frontier,'" *Science* 191 (January 9): 41–47.

England, J. Merton. 1970. "Interesting Times—The NSF since 1960," *Mosaic* 1: 3–7.

Epstein, Steven. 1991. "Democratic Science? AIDS Activism and the Contested Construction of Knowledge," *Socialist Review* 91 (2): 35–64.

Ergas, Henry. 1987. "Does Technology Policy Matter?" In Bruce R. Guile and Harvey Brooks (eds.), *Technology and Global Industry* (Washington, D.C.: National Academy Press), 191–245.

Etzkowitz, Henry. 1983. "Entrepreneurial Scientists and Entrepreneurial Universities in American Academic Science," *Minerva* 21: 198–233.

Evans, Peter B., Dietrich Rueschemeyer, and Theda Skocpol. 1985. "On the

Road Toward a More Adequate Understanding of the State." In Peter B. Evans et al. (eds.), *Bringing the State Back In* (New York: Cambridge University Press), 347–66.

Feister, Irving. 1947. "Wanted: An Integrated Science Foundation," *Nation* 165 (October 25): 456–57.

Fisher, Donald. 1980. "American Philanthropy and the Social Sciences: The Reproduction of a Conservative Ideology," *Sociological Review* 28(2): 277–315.

Forman, Paul. 1987. "Behind Quantum Electronics: National Security as the Basis for Physical Research in the United States, 1940–1960," *Historical Studies in the Physical and Biological Sciences* 18(1): 149–229.

Fortune. 1946. "The Great Science Debate," *Fortune* (June): 116–20, 236, 239–40, 242, 245.

Friedman, Robert S. and Renee C. Friedman. 1988. "Science American Style: Three Cases in Academe," *Policy Studies Journal* 17(1): 41–61.

Fuchs, Stephan. 1992. *The Professional Quest for Truth: A Social Theory of Science and Knowledge* (Albany: State University of New York Press.

Furedy, John J. 1987. "Melding Capitalist versus Socialist Models of Fostering Scientific Excellence." In Douglas Jackson and J. Philippe Ruhton (eds.), *Scientific Excellence: Origins and Assessment* (Newbury Park, CA: Sage), 284–306.

Gable, Richard W. 1953. "NAM: Influential Lobby or Kiss of Death," *Journal of Politics* 15(2): 254–73.

Geiger, Roger L. 1993. *Research and Relevant Knowledge: American Research Universities since World War II* (New York: Oxford University Press).

Geiger, Roger L. 1986. *To Advance Knowledge: The Growth of American Research Universities, 1900–1940* (New York: Oxford University Press).

Genuth, Joel. 1988. "Microwave Radar, the Atomic Bomb, and the Background to U.S. Research Priorities in World War II," *Science, Technology, and Human Values* 13(3–4): 276–89.

Genuth, Joel. 1987. "Groping towards Science Policy in the United States in the 1930s," *Minerva* 25: 238–68.

Gibbs, Lois Marie. 1982. *Love Canal: My Story* (Albany, NY: SUNY Press).

Gilbert, Jess and Carolyn Howe. 1990. "Beyond 'State vs. Society': Theories of the State and New Deal Agricultural Policies," *American Sociological Review* 56: 204–20.

Gilpin, Robert G., Jr. 1975. "Science, Technology, and French Independence." In T. Dixon Long and Christopher Wright (eds.), *Science Policies of Industrial Nations: Case Studies of the United States, Soviet Union, United Kingdom, France, Japan, and Sweden* (New York: Praeger), 110–32.

Gilpin, Robert. 1968. *France in the Age of the Scientific State* (Princeton, NJ: Princeton University Press).

Glass, Bentley. 1960. "The Academic Scientist: 1940–1960," *Science* 132: (September 2): 598–603.

Gold, David, Clarence Lo, and Erik Wright. 1975. "Recent Developments in Marxist Theories of the State," *Monthly Review* (October): 29–43; (November): 36–51.

Gourevitch, Peter. 1986. *Politics in Hard Times.* (Ithaca, NY: Cornell University Press).

Graham, Margaret B. W. 1985. "Industrial Research in the Age of Big Science," *Research on Technological Innovation, Management and Policy* 2: 47–79.

Greenberg, Daniel S. 1967. *The Politics of Pure Science* (New York: New American Library).

Greenberg, Daniel S. 1963. "Civilian Technology: Program to Boost Industrial Research Heavily Slashed in House," *Science* 140 (June 28): 1380–83.

Greenhouse, Steven. 1987. "When the World's Growth Slows," *New York Times,* 27 December, section 3, p. 1.

Gruber, Carol. 1975. *Mars and Minerva: World War I and the Uses of Higher Education in America* (Baton Rouge: Louisiana State University).

Gummett, Philip J. and Geoffrey L. Price. 1977. "An Approach to the Central Planning of British Science: The Formation of the Advisory Council on Scientific Policy," *Minerva* 25(2): 119–43.

Hagstrom, Warren. 1965. *The Scientific Community* (London: Feffer & Simons).

Hall, Peter. 1986. *Governing the Economy: The Politics of State Intervention in Britain and France* (Cambridge, England: Polity Press).

Hamilton, David P. 1992. "National Science Board Sounds Wake-Up Call," *Science* 257 (August 21): 1039.

Hamilton, David P. 1991. "Industrial R&D Plea," *Science* 253 (September 20): 1350.

Hamilton, David P. 1990. "Technology Policy: Congress Takes the Reins," *Science* 250 (November 9): 747.

Hoch, Paul K. 1988. "The Crystallization of a Strategic Alliance: The American Physics Elite and the Military in the 1940s." In E. Mendelsohn, M. R. Smith, and P. Weingart, *Science, Technology, and the Military,* vol. 12 (Dordrecht, The Netherlands: Kluwer Academic Publishers) 87–116.

Hodes, Elizabeth. 1982. "Precedents for Social Responsibility among Scientists: The American Association of Scientific Workers and the Federation of American Scientists, 1938–1948." Unpublished Ph.D. dissertation, University of California, Santa Barbara.

Holden, Constance. 1993. "Clinton's New Policy: More Is Less," *Science* 262 (December 10): 1643.

Holton, Gerald. 1979. "From the Endless Frontier to the Ideology of Limits." In

Gerald Holton and Robert Morison (eds.), *Limits of Scientific Inquiry* (New York: Norton), 227–41.

Hooks, Gregory. 1991. *Forging the Military-Industrial Complex: World War II's Battle of the Potomac* (Chicago: University of Illinois Press).

Hooks, Gregory. 1990a. "From an Autonomous to a Captured State Agency: The Decline of the New Deal in Agriculture," *American Sociological Review* 55(1): 29–43.

Hooks, Gregory. 1990b. "The Rise of the Pentagon and U.S. State-Building: The Defense Program as Industrial Policy," *American Journal of Sociology* 96: 358–404.

Humphrey, Hubert H. 1960. "The Need for a Department of Science," *Annals of the American Academy of Political and Social Sciences* 327 (January): 27–35.

Hunt, James B. 1982. "State Involvement in Science and Technology," *Science* 215 (February 5): 4533.

Ikenberry, John G. 1988. "Conclusion: An Institutional Approach to American Foreign Economic Policy," *International Organization* 42(1): 219–43.

Inman, B. R. and Daniel F. Burton. 1991. "Rx: A Technology Policy," *New York Times*, January 17, p. A19. (advertisement)

Jessop, Bob. 1982. *The Capitalist State* (New York: New York University Press).

Joint Economic Committee. 1992. *Technology and Economic Performance*. Hearing before the Joint Economic Committee, Congress of the United States, 102nd Congress, First Session, September 12 (Washington, D.C.: Government Printing Office).

Jones, Kenneth M. 1976. "The Endless Frontier," *Prologue* 8 (spring): 35–46.

Jones, Kenneth M. 1975. "Science, Scientists, and Americans: Images of Science and the Formation of Federal Science Policy." Unpublished Ph.D. dissertation, Cornell University, Ithaca, New York.

Kaempffert, Waldemaer. 1943. "The Case for Planned Research," *American Mercury* 57 (October): 442–47.

Kantrowitz, Arthur. 1975. "Controlling Technology Democratically," *American Scientist* 63 (September–October): 505–09.

Kargon, Robert and Elizabeth Hodes. 1985. "Karl Compton, Isaiah Bowman, and the Politics of Science in the Great Depression," *Isis* 76: 301–18.

Katz, James E. 1980. "Organizational Structure and Advisory Effectiveness: The Office of Science and Technology Policy." In William T. Golden (ed.), *Science Advice to the President* (New York: Pergamon), 229–44.

Katzenstein, Peter J. (ed.). 1978. *Between Power and Plenty: Foreign Economic Policies of Advanced Industrial States* (Madison: University of Wisconsin Press), 295–336.

Kevles, Daniel J. 1990. "Principles and Politics in Federal R&D Policy, 1945–1990: An Appreciation of the Bush Report." In Vannevar Bush, *Science—The*

Endless Frontier [40th anniversary reissue] (Washington, D.C.: National Science Foundation), ix–xxxiii.

Kevles, Daniel J. 1988a. "American Science." In Nathan O. Hatch (ed.), *The Professions in American History* (Notre Dame, IN: Notre Dame University Press), 107–25.

Kevles, Daniel J. 1988b. "Cold War and Hot Physics: Reflections on Science, Security and the American State." *Humanities Working Paper* 135. (Pasadena, California Institute of Technology).

Kevles, Daniel J. 1987. *The Physicists: The History of a Scientific Community in Modern America* (Cambridge, MA: Harvard University Press).

Kevles, Daniel J. 1978. "Notes on the Politics of American Science: Commentary on Papers by Alice Kimball Smith and Dorothy Nelkin," *Science, Technology, and Human Values* 24: 40–44.

Kevles, Daniel J. 1977. "The National Science Foundation and the Debate over Postwar Research Policy," *Isis* 68: 5–26.

Kevles, Daniel J. 1975. "Scientists, the Military, and the Control of Postwar Defense Research: The Case of the Research Board for National Security," *Technology and Culture* 16: 20–47.

Kevles, Daniel J. 1974. "FDR's Science Policy," *Science* 183: 798–800.

Kilgore, Harley. 1943. "Discussion: The Science Mobilization Bill," *Science* 98 (2537): 151–52.

Kleinman, Daniel Lee. 1991. "Conceptualizing the Politics of Science: A Response to Cambrosio, Limoges and Pronovost," *Social Studies of Science* 21(4): 769–74.

Kleinman, Daniel Lee and Jack R. Kloppenburg Jr. 1991. "Aiming for the Discursive High Ground: Monsanto and the Biotechnology Controversy," *Sociological Forum* 6: 427–47.

Kleppner, Daniel. 1991. "The Ending Frontier: The Lederman Report and Its Critics," *Issues in Science and Technology* (spring): 32–37.

Kloppenburg, Jack R. Jr. 1988. *First the Seed: The Political Economy of Plant Biotechnology, 1492–2000* (New York: Cambridge University Press).

Knorr-Cetina, Karin. 1983. "The Ethnographic Study of Scientific Work—Towards a Constructivist Interpretation of Science." In Karin Knorr Cetina and Michael Mulkay (eds.), *Science Observed: Perspectives on the Social Study of Science* (London: Sage), 115–40.

Kohler, Robert E. 1990. *Partners in Science: Foundations and Natural Scientists, 1900–1945* (Chicago: University of Chicago Press).

Kohler, Robert E. 1987. "Science, Foundations, and American Universities in the 1920s," *Osiris* 2(3): 135–64.

Kohler, Robert E. 1979. "Warren Weaver and the Rockefeller Foundation Program in Molecular Biology: A Case Study in the Management of Science." In

Nathan Reingold (ed.), *The Sciences in the American Context: New Perspectives* (Washington, D.C.: Smithsonian Institution Press), 249–93.

Koistinen, Paul A. C. 1972. "The Military-Industrial Complex in Historical Perspective: The Interwar Years." In Carroll Pursell (ed.), *The Military-Industrial Complex* (New York: Harper & Row), 31–50.

Koshland, Daniel E. 1988. "Setting Priorities in Science," *Science* 240: (May 20): 4855.

Koshland, Daniel E. 1985. "A Department of Science?," *Science* 227 (February 8): 4687.

Krasner, Stephen. 1984. "Approaches to the State: Alternative Conceptions and Historical Dynamics," *Comparative Politics* 16 (January): 223–46.

Krasner, Stephen D. 1978. *Defending the National Interest* (Princeton, NJ: Princeton University Press).

Krimsky, Sheldon. 1982. *Genetic Alchemy: The Social History of the Recombinant DNA Controversy* (Cambridge, MA: MIT Press).

Kuznick, Peter J. 1987. *Beyond the Laboratory: Scientists as Political Activists in 1930s America* (Chicago: University of Chicago Press).

Kwa, Chunglin. 1987. "Representation of Nature Mediating between Ecology and Science Policy: The Case of the International Biological Programme," *Social Studies of Sciences* 17: 413–42.

Lambright, W. Henry and Dianne Rahm (eds.). 1992. *Technology and U.S. Competitiveness: An Institutional Focus* (New York: Greenwood).

Lapp, Ralph E. 1965. *The New Priesthood: The Scientific Elite and the Uses of Power* (New York: Harper & Row).

Larson, Magali Sarfatti. 1984. "The Production of Expertise and the Constitution of Expert Power." In Thomas Haskell (ed.), *The Authority of Experts* (Bloomington: Indiana University Press), 28–80.

Larson, Magali Sarfatti. 1977. *The Rise of Professionalization: A Sociological Analysis* (Berkeley: University of California Press).

Lasby, Clarence. 1966. "Science and the Military." In David Van Tassel and Michael Hall (eds.), *Science and Society in the United States* (Homewood, IL: Dorsey), 251–82.

Latour, Bruno. 1987. *Science in Action: How to Follow Scientists and Engineers through Society* (Cambridge, MA: Harvard University Press).

Latour, Bruno and Steve Woolgar. 1979. *Laboratory Life: The Social Construction of Scientific Facts* (Beverly Hills: Sage).

Lederman, Leon. 1991. *Science: The End of the Frontier?* (Washington, D.C.: American Association for the Advancement of Science).

Lederman, Leonard. 1987. "Science and Technology Policies and Priorities: A Comparative Analysis," *Science* 237 (September 4): 1125–33.

Lederman, Leonard L. 1985. "Science and Technology in Europe: A Survey," *Science and Public Policy* 12(3): 131–43.

Lederman, Leonard L., Rolf Lehming, and Jennifer S. Bond. 1986. "Research Policies and Strategies in Six Countries: A Comparative Analysis," *Science and Public Policy* 13(2): 67–76.

Lepkowski, Wil. 1991. "Critical Technologies: White House Stalls Institute's Creation," *Chemical and Engineering News* (September 16): 4–5.

Leslie, Stuart W. 1993. *The Cold War and American Science: The Military-Industrial-Academic Complex at MIT and Stanford* (New York: Columbia University Press).

Leslie, Stuart W. 1987. "Playing the Education Game to Win: The Military and Interdisciplinary Research at Stanford," *Historical Studies in the Physical Sciences* 18(1): 55–88.

Lessing, Lawrence P. 1954. "The National Science Foundation Takes Stock," *Scientific American* (March): 29–33.

Leuchtenburg, William E. 1963. *Franklin D. Roosevelt and the New Deal* (New York: Harper & Row).

Levine, Rhonda. 1988. *Class Struggle and the New Deal: Industrial Labor, Industrial Capital and the State* (Lawrence: University of Kansas Press).

Lieberson, Stanley. 1992. "Einstein, Renoir, and Greeley: Some Thoughts About Evidence in Sociology." *American Sociological Review* 57: 1–15.

Lindberg, Leon. 1985. "Political Economy, Economic Governance, and the Coordination of Economic Activities," *Wissenschaftskolleg Jahrbuch* (1984/85): 241–55.

Lindberg, Leon. 1982. "The Problems of Economic Theory in Explaining Economic Performance," *Annals of the American Academy of Political and Social Sciences* 459 (January): 14–27.

Lindberg, Leon, Fritz Scharpt, and Gunter Engelhardt. 1987. "Economic Policy Research: Challenges and a New Agenda." In Meinolf Dierkes, Hans Weiler, and Ariane Berthorn Antal (eds.), *Comparative Policy Research: Learning from Experience* (New York: St. Martins), 347–78.

Lomask, Milton. 1976. *A Minor Miracle: An Informal History of the National Science Foundation* (Washington, D.C.: National Science Foundation).

Lomask, Milton. 1973. "Historical Footnote," *Science* 182 (October 12): 116.

Long, T. Dixon. 1975. "The Dynamics of Japanese Science Policy." In T. Dixon Long and Christopher Wright (eds.), *Science Policies of Industrial Nations: Case Studies of the United States, Soviet Union, United Kingdom, France, Japan, and Sweden* (New York: Praeger), 133–68.

Lowi, Theodore J. 1967. "Party, Policy, and Constitution in America." In William Nisbet Chambers and Walter Dean Burnham (eds.), *The American Party Systems* (New York: Oxford University Press), 238–76.

MacKenzie, Donald and Graham Spinardi. 1988a: "The Shaping of Nuclear Weapon System Technology: U.S. Fleet Ballistic Missile Guidance and Navigation. I: From Polaris to Poseidon," *Social Studies of Science* 18: 419–63.

MacKenzie, Donald and Graham Spinardi. 1988b. "The Shaping of Nuclear Weapon System Technology: U.S. Fleet Ballistic Missile Guidance and Navigation. II: 'Going for Broke'—The Path to Trident II," *Social Studies of Science* 18: 581–624.

Maddox, Robert F. 1981. *The Senatorial Career of Harley Martin Kilgore* (New York: Garland).

Maddox, Robert F. 1979. "The Politics of World War II Science: Senator Harley M. Kilgore and the Legislative Origins of the National Science Foundation," *West Virginia History* 41(1): 20–39.

Mann, Michael. 1988. *States, War and Capitalism: Studies in Political Sociology* (New York: Basil Blackwell).

Mansfield, Edwin. 1988. "Industrial Innovation in Japan and the United States," *Science* 241 (September 30): 1769–74.

Markusen, Ann and Joel Yudken. 1992. "Building a New Economic Order," *Technology Review* (April): 23–30.

Marshall, Eliot. 1993a. "R&D Policy that Emphasizes the 'D,'" *Science* 259 (March 26): 1816–19.

Marshall, Eliot. 1993b. "Swords to Plowshares Plan Boosts R&D," *Science* 259 (March 19): 1690.

Marshall, Eliot. 1992. "Space Scientists Heed Call to Set Priorities," *Science* 255 (January 31): 527–28.

Marshall, Eliot. 1991. "U.S. Technology Strategy Emerges," *Science* 252 (April 5): 20–24.

Martin, Andrew. 1985. "Wages, Profits, and Investment in Sweden." In Leon Lindberg and Charles Maier (eds.), *The Politics of Inflation and Economic Stagnation* (Washington, D.C.: Brookings Institution), 403–66.

Mazuzan, George T. 1987. *The National Science Foundation: A Brief History* (Washington, D.C.: National Science Foundation).

McClenahen, John S. 1990. "Standards Czar Eyes Technology," *Industry Week* (July 2): 51.

McCune, Robert P. 1971. *Origins and Development of the National Science Foundation and Its Division of Social Sciences, 1945–1961*. Ph.D. dissertation, Ball State University; University Microfilms, Ann Arbor Michigan.

McDougall, Walter A. 1985. . . . *The Heavens and the Earth: A Political History of the Space Age* (New York: Basic Books).

McLauchlan, Gregory. 1989. "World War, the Advent of Nuclear Weapons, and Global Expansion of the National Security State." In Robert K. Schaeffer (ed.), *War in the World-System* (New York: Greenwood), 83–97.

Meigs, Montgomery Cunningham. 1982. "Managing Uncertainty: Vannevar Bush, James B. Conant and the Development of the Atomic Bomb." Unpublished Ph.D. dissertation, University of Wisconsin–Madison.

Merton, Robert K. 1973. *The Sociology of Science: Theoretical and Empirical Investigations* (Chicago: University of Chicago Press).

Mervis, Jeffrey. 1993. "Lane's Strategy on Strategic Research," *Science* 262 (November 12): 983.

Mervis, Jeffrey, Christopher Anderson, and Eliot Marshall. 1993. "Better for Science than Expected," *Science* 262 (November 5): 836–39.

Miliband, Ralph. 1983. "State Power and Class Interests," *New Left Review* 138 (March–April): 57–68.

Miliband, Ralph. 1969. *The State in Capitalist Society* (New York: Basic Books).

Mills, C. Wright. 1956. *The Power Elite* (London: Oxford University Press).

Moore, John Robert. 1967. "The Conservative Coalition in the United States Senate, 1942–1945," *Journal of Southern History* 33 (August): 368–76.

Mosaic. 1970. "A Visit with Vannevar Bush," *Mosaic* 1: 9–12.

Mowery, David C. 1993. "Whither DARPA?" *Issues in Science and Technology* (summer): 6.

Mowery, David C. 1983. "The Relationship between Intrafirm and Contractual Forms of Industrial Research in American Manufacturing, 1900–1940," *Explorations in Economic History* 20: 351–74.

Mulkay, Michael. 1980. "Interpretation and the Use of Rules: The Case of Norms in Science." In Thomas Gieryn (ed.), *Science and Social Structure: A Festschrift for Robert K. Merton* (New York: New York Academy of Science), 111–25.

National Academy of Sciences, National Academy of Engineering, and Institute of Medicine. 1993. *Science, Technology, and the Federal Government: National Goals for a New Era* (Washington, D.C.: National Academy Press).

National Research Council. 1940. *Research a National Resource. II. Industrial Research* (Washington, D.C.: U.S. Government Printing Office).

Needell, Allan A. 1987. "Preparing for the Space Age: University-Based Research, 1946–1957," *Historical Studies in the Physical and Biological Sciences* 18(1): 89–109.

Nelkin, Dorothy. 1984. "Science and Technology Policy and the Democratic Process." In James C. Petersen (ed.), *Citizen Participation in Science Policy* (Amherst, MA: University of Massachusetts Press), 18–39.

Nelson, William Richard. 1965. "Case Study of a Pressure Group: The Atomic Scientists." Unpublished Ph.D. dissertation, University of Colorado, Department of Political Science.

Nettl, J. P. 1968. "The State as a Conceptual Variable," *World Politics* 20: 559–92.

New Republic. 1947. "The Ivy League Lobby," *New Republic* 117 (August 4): 10.

New York Times. 1992. "Industrial Policy as Sloppy Slogan," February 12, p. A12.

New York Times. 1945. "Truman Aid Asked for Magnuson Bill," November 27, p. 15.

Nichols, David. 1974. "The Associational Interest Groups of American Science." In Albert H. Teich (ed.), *Scientists and Public Affairs* (Cambridge, MA: MIT Press), 123–70.

Nielsen, Waldemar A. 1989. *The Golden Donors: A New Anatomy of the Great Foundations* (New York: Truman Talley).

Nielsen, Waldemar A. 1972. *The Big Foundations: A Twentieth Century Fund Study* (New York: Columbia University Press).

Noble, David. 1984. *Forces of Production: A Social History of Industrial Automation* (New York: Knopf).

Noble, David. 1983. "Academia Incorporated," *Science for the People* (January/February): 7–11, 50–52.

Noble, David. 1977. *America by Design: Science, Technology, and the Rise of Corporate Capitalism* (New York: Oxford University Press).

Norman, Colin. 1992. "Science Budget: Selective Growth," *Science* 255 (February 7): 672–75.

Norman, Colin. 1990. "Defense Research after the Cold War," *Science* 247 (January 19): 272–73.

Norman, Colin. 1988. "Technology Legislation Previewed," *Science* 242 (November 11): 861.

Office of Science and Technology Policy. 1992. "Technology for a Productive America—Fact Sheet." Unpublished document (Washington, D.C.: Executive Office of the President, OSTP).

Office of Science and Technology Policy. 1990. *U.S. Technology Policy* (Washington, D.C.: Executive Office of the President, OSTP).

Office of Technology Assessment. 1991. *Federally Funded Research: Decisions for a Decade,* summary and complete report (Washington, D.C.: U.S. Government Printing Office).

Orloff, Ann Shola. 1988. "The Political Origins of America's Belated Welfare State." In Margaret Weir, Ann Orloff, and Theda Skocpol (eds.), *The Politics of Social Policy in the United States* (Princeton, NJ: Princeton University Press), 37–80.

Orloff, Ann Shola, and Theda Skocpol. 1984. " 'Why Not Equal Protection?': Explaining the Politics of Social Spending in Britain, 1900–1911 and the United States, 1880s–1920," *American Sociological Review* 49: 726–50.

Owens, Larry. 1987. "Straight-Thinking: Vannevar Bush and the Culture of American Engineering." Unpublished Ph.D. dissertation, Princeton University, Princeton, New Jersey.

Palca, Joseph. 1992a. "Congress Queries Hallowed Principles," *Science* 257 (September 18): 1620.

Palca, Joseph. 1992b. "Massey Seeks to Broaden NSF's Role," *Science* 257 (August 21): 1035.

Palca, Joseph. 1991. "OTA Challenges Dogma on Research Funding," *Science* 251 (March 29): 1555.

Palmer, Archie. 1948. "Industry Supported University Research," *Chemical and Engineering News* 26 (July 12): 2042–45.

Parsons, Talcott. 1946. "The Science Legislation and the Role of the Social Sciences," *American Sociological Review* 11 (December): 653–66.

Penick, James L., Carroll W. Pursell Jr., Morgan B. Sherwood, and Donald C. Swain (eds.). 1972 [1965]. *The Politics of American Science* (Cambridge, MA: MIT Press).

Perazich, George and Philip M. Field. 1940. *Industrial Research and Changing Technology* (Philadelphia, PA: Works Projects Administration).

Perez, Carlota. 1986. "Structural Changes and the Assimilation of New Technologies in the Economic and Social System." In Christopher Freeman (ed.), *Design, Innovation and Long Cycles in Economic Development* (New York: St. Martin's), 27–47.

Petersen, James C. 1984. "Citizen Participation in Science Policy." In James C. Petersen (ed.), *Citizen Participation in Science Policy* (Amherst: University of Massachusetts Press), 1–17.

Piore, Michael and Charles Sabel. 1984. *The Second Industrial Divide: Possibilities for Prosperity* (New York: Basic Books).

Pious, Richard M. 1979. *The American Presidency* (New York: Basic Books).

Polanyi, Michael. 1962. "The Republic of Science," *Minerva* 1 (autumn): 54–73.

Polanyi, Michael. 1951. *The Logic of Liberty: Reflections and Rejoinders* (Chicago: University of Chicago Press).

Pollack, Andrew. 1990. "High Tech Business Loses a Friend at the Pentagon," *New York Times,* April 29, section 4, p. 5.

Pollack, Andrew. 1989a. "America's Answer to MITI," *New York Times,* March 5, Business, p. 1.

Pollack, Andrew. 1989b. "Panel Asks Strong U.S. Push to Develop Superconductors," *New York Times,* January 4, section 1, pp. 1, 27.

Powers, Phillip. 1947. "The Organization for Science in the Federal Government, *Bulletin of Atomic Scientists* 3 (April/May): 122–23, 126.

Prechel, Harland. 1990. "Steel and the State: Industry Politics and Business Policy Formation, 1940–1989," *American Sociological Review* 55(5):

President's Scientific Research Board (John R. Steelman, Chair). 1947. *Science and Public Policy,* 5 vols. (Washington, D.C.: U.S. Government Printing Office).

Press, Frank. 1988. "The Dilemma of the Golden Age," *Science, Technology, and Human Values* 13: 224–31.

Price, Don K. 1979. "Endless Frontier or Bureaucratic Morass." In Gerald Holton and Robert Morison (eds.), *Limits of Scientific Inquiry* (New York: Norton), 75–92.

Price, Don K. 1965. *The Scientific Estate* (Cambridge, MA: Belknap).

Pursell, Carroll. 1985. "The Search for a Department of Science: An Historical Overview." Unpublished paper prepared for the National Science Foundation, Washington, D.C.

Pursell, Carroll. 1979a. "Government and Technology in the Great Depression," *Technology and Culture* 20(1): 162–74.

Pursell, Carroll. 1979b. "Science Agencies in World War II: The OSRD and its Challengers." In Nathan Reingold (ed.), *The Sciences in the American Context: New Perspectives* (Washington, D.C.: Smithsonian Institution Press), 359–78.

Pursell, Carroll. 1976. "Alternative American Science Policies during World War II." In James E. O'Neill and Robert W. Krauskopf (eds.), *World War Two: An Account of Its Documents* (Washington, D.C.: Howard University Press), 151–62.

Pursell, Carroll. 1971. "American Science Policy during World War II," *International Congress on the History of Science* 13(2): 274–78.

Pursell, Carroll. 1966. "Science and Government Agencies." In David Van Tassel and Michael Hall (eds.), *Science and Society in the United States* (Homewood, IL: Dorsey), 223–49.

Pursell, Carroll. 1965. "Anatomy of a Failure: The Science Advisory Board, 1933–1935," *Proceedings of the American Philosophical Society* 109(6): 342–51.

Quadagno, Jill. 1991. "Who Rules Sociology Now?" *Contemporary Sociology* 20(4): 563–65.

Quadagno, Jill and Madonna Harrington Meyer. 1989. "Organized Labor, State Structures, and Social Policy Development: A Case Study of Old Age Assistance in Ohio, 1916–1940," *Social Problems* 36(2): 181–96.

Rabinowitch, Eugene. 1946. "Science, A Branch of the Military?" *Bulletin of Atomic Scientists* 2 (November 1): 1.

Rankin, William L., Stanley M. Nealey, and Barbara Desow Melber. 1984. "Overview of National Attitudes toward Nuclear Energy: A Longitudinal Analysis." In William R. Freudenburg and Eugene A. Rosa (eds.), *Public Reaction to Nuclear Power* (Boulder, CO: Westview), 41–68.

Redmond, Kent. 1968. "World War II, a Watershed in the Role of the National Government in the Advancement of Science and Technology." In Charles Angoff (ed.), *The Humanities in the Age of Science* (Rutherford, NJ: Fairleigh Dickinson University Press), 166–80.

Reingold, Nathan. 1987. "Vannevar Bush's New Deal for Research: Or the Triumph of the Old Order," *Historical Studies in the Physical Sciences* 17(2): 299–344.

Reingold, Nathan. 1977. "The Case of the Disappearing Laboratory," *American Quarterly* 29: 79–101.

Research Technology Management. 1991. "First Grants for Key Industrial Technologies" 34: 3.

Restivo, Sal. 1988. "Modern Science as a Social Problem," *Social Problems* 35(3): 206–25.

Rhodes, Frank H. T. 1988. "A System to Set Science Priorities," *Technology Review* 91: 21–25.

Rhodes, Richard. 1986. *The Making of the Atomic Bomb* (New York: Simon & Schuster).

Roland, Alex. 1985a. *Model Research: The National Advisory Committee on Aeronautics, 1915–1958* (Washington, D.C.: National Aeronautics and Space Administration).

Roland, Alex. 1985b. "Science and War," *Osiris* 2(1): 247–72.

Ronayne, J. 1984. *Science in Government: A Review of the Principles and Practice of Science Policy* (London: Edward Arnold).

Rosenberg, Nathan and Claudio R. Frischtak. 1986. "Technological Innovation and Long Waves." In Christopher Freeman (ed.), *Design, Innovation and Long Cycles in Economic Development* (New York: St. Martin's), 5–26.

Rosenzweig, Robert M. 1988. "Thinking about Less," *Science, Technology, and Human Values* 13 (summer and autumn): 219–23.

Rossiter, Margaret W. 1980. "American Science in the 1970s [a Review of Nathan Reingold (ed.), *The Sciences in the American Context*]," *Reviews in American History* (December): 547–52.

Roth, William V. 1993. "Whither DARPA?" *Issues in Science and Technology* (summer): 5.

Rothwell, Roy and Walter Zegveld. 1985. *Reindustrialization and Technology* (Essex, England: Longman).

Roush, Wade. 1990. "Science and Technology in the 101st Congress," *Technology Review* (November/December): 59–69.

Rowan, Carl Milton. 1985. "Politics and Pure Research: The Origins of the National Science Foundation, 1942–1954." Unpublished Ph.D. dissertation, Miami University, Oxford, OH.

Salzman, Harold and G. William Domhoff. 1980. "The Corporate Community and Government: Do They Interlock?" In G. William Domhoff (ed.), *Power Structure Research* (Beverly Hills: Sage), 227–54.

Sapolsky, Harvey. 1990. *Science and the Navy: The History of the Office of Naval Research* (Princeton, NJ: Princeton University Press).

Sapolsky, Harvey. 1979. "Academic Science and The Military: The Years Since the Second World War." In Nathan Reingold (ed.), *The Sciences in the American*

Context: New Perspectives (Washington, D.C.: Smithsonian Institution Press), 379–99.

Schaffter, Dorothy. 1969. *The National Science Foundation* (New York: Praeger).

Schmitt, Roland W. 1986. "Improving International Competitiveness: The Federal Role," *Technology in Society* 8: 129–36.

Schooler, Dean Jr. 1971. *Science, Scientists, and Public Policy* (New York: Free Press).

Schriftgiesser, Karl. 1951. *The Lobbyists: The Art and Business of Influencing Lawmakers* (Boston: Little, Brown).

Science. 1991. "Going Critical over CTI," *Science* 253 (September 20): 1343.

Science. 1990. "Bad News from the Competitiveness Front," *Science* 248 (June 8): 1185.

Science. 1989. "Glenn Proposes Civilian DARPA," *Science* 245: (December 8): 1252.

Science. 1946. "Obituary: NSF, 1946," *Science* 104 (August 2): 97–98.

Science. 1943. "The Mobilization of Science" [text of S. 702], *Science* 97 (May 7): 407–12.

Sewell, William H. Forthcoming. "Three Temporalities: Toward an Eventful Sociology." In Terrence J. McDonald (ed.), *The Historic Turn in the Human Sciences* (Ann Arbor: University of Michigan Press).

Shapley, Deborah and Rustum Roy. 1985. *Lost at the Frontier: U.S. Science and Technology Policy Adrift* (Philadelphia: ISI Press).

Shefter, Martin. 1978. "Party, Bureaucracy, and Political Change in the United States." In Louis Maisel and Joseph Cooper (eds.), *Political Parties: Development and Decay* (Beverly Hills: Sage), 211–66.

Shefter, Martin. 1977. "Party and Patronage: Germany, England, and Italy," *Politics and Society* 7(4): 403–52.

Sherwood, Morgan. 1968. "Federal Policy for Basic Research: Presidential Staff and the National Science Foundation, 1950–1956," *Journal of American History* 55: 599–615.

Shonfield, Andrew. 1965. *Modern Capitalism: The Changing Balance of Public and Private Power* (New York: Oxford University Press).

Skocpol, Theda. 1985. "Bringing the State Back In: Strategies of Analysis in Current Research." In Peter Evans, Dietrich Rueschemeyer, and Theda Skocpol (eds.), *Bringing the State Back In* (New York: Cambridge University Press), 3–37.

Skocpol, Theda. 1984. "Emerging Agendas and Recurrent Strategies in Historical Sociology." In Theda Skocpol (ed.), *Vision and Method in Historical Sociology* (New York: Cambridge University Press), 356–91.

Skocpol, Theda. 1980. "Political Responses to Capitalist Crisis: Neo-Marxist

Theories of the State and the Case of the New Deal," *Politics and Society* 10(2): 155–201.

Skocpol, Theda and Kenneth Finegold. 1982. "State Capacity and Economic Intervention in the Early New Deal," *Political Science Quarterly* 97: 255–78.

Skowronek, Stephen. 1982. *Building the New American State: The Expansion of National Administrative Capacities, 1877–1920* (New York: Cambridge University Press).

Smith, Alice Kimball. 1978. "Scientists and the Public Interest—1945–46," *Science, Technology, and Human Values* 24: 24–32.

Smith, Alice Kimball. 1970. *A Peril and a Hope: The Scientists' Movement in America: 1945–47* (Cambridge, MA: MIT Press).

Smith, Bruce L. R. 1990. *American Science Policy since World War II* (Washington, D.C.: Brookings Institution).

Smith, John K. and David A. Hounshell. 1985. "Wallace H. Carothers and Fundamental Research at Du Pont," *Science* 229 (August 2): 436–42.

Steel. 1946. "Kilgore Injects Conservatism into New Science Program Bill," *Steel* 118 (January 21): 58–60.

Stewart, Irvin. 1948. *Organizing Scientific Research for the War: The Administrative History of the Office of Scientific Research and Development* (Boston, MA: Little, Brown).

Stone, Richard. 1993. "Congress Gunning for Science Cuts," *Science* 262 (November 12): 979.

Strickland, Stephen P. 1972. *Politics, Science, and the Dread Disease* (Cambridge, MA: Harvard University Press).

Strong, Elizabeth. 1982. "Science and the Early New Deal," *Synthesis* 5(2): 44–63.

Study Group, Washington Association of Scientists. 1947. "Toward a National Science Policy," *Science* 106 (October 24): 385–87.

Sturchio, Jeffrey L. 1984. "Chemistry and Corporate Strategy at Du Pont," *Research Management* 27(1): 10–15.

Swain, Donald. 1962. "The Rise of a Research Empire: NIH, 1930–1950," *Science* 138: 1233–37.

Swan, John P. 1988. *Academic Scientists and the Pharmaceutical Industry: Cooperative Research in Twentieth Century America* (Baltimore: Johns Hopkins University Press).

Szelenyi, Ivan and Bill Martin. 1988. "The Three Waves of New Class Theories," *Theory and Society* 17: 645–67.

Technology Policy Task Force, Committee on Science, Space and Technology, U.S. House of Representatives. 1988. *Technology Policy and Its Effects on the National Economy* (Washington, D.C.: U.S. Government Printing Office).

Tedlow, Richard S. 1976. "The National Association of Manufacturers and Public Relations during the New Deal," *Business History Review* 50(1): 25–55.

Teich, Albert. 1990. "Scientists and Public Officials Must Pursue Collaboration to Set Research Priorities," *Scientist* (February 5): 10, 19.

Tilly, Charles. 1990. *Coercion, Capital, and European States, AD 990–1990* (Cambridge, MA: Basil Blackwell).

Tilly, Charles. 1975. "Reflections on the History of European State-Making." In Charles Tilly (ed.), *The Formation of National States in Western Europe* (Princeton, NJ: Princeton University Press), 2–83.

Tour, Sam. 1948. "Industry Supported University Research," *Chemical and Engineering News* 26 (July 12): 2046–48.

U.S. House of Representatives, Committee on Science, Space, and Technology. 1992. *Report of the Task Force on the Health of Research,* 102nd Congress, Second Session (Washington, D.C.: U.S. Government Printing Office).

U.S. House of Representatives, Committee on Science, Space, and Technology, Subcommittee on Science, Research and Technology. 1987. *The Role of Science and Technology in Competitiveness* (hearings). 100th Congress, First Session, April 28, 29, 30 (Washington, D.C.: U.S. Government Printing Office).

U.S. House of Representatives, Committee on Science and Technology, Task Force on Science Policy [cited as U.S. House Task Force on Science Policy]. 1986a. *Science Policy Study Background Report Number 1: A History of Science Policy in the United States, 1940–1985.* 99th Congress, Second Session (Washington, D.C.: U.S. Government Printing Office).

U.S. House of Representatives, Committee on Science and Technology, Task Force on Science Policy [cited as U.S. House Task Force on Science Policy]. 1986b. *Science Policy Study Background Report No. 8: Science Support by the Department of Defense,* 99th Congress, Second Session (Washington, D.C.: U.S. Government Printing Office).

U.S. House of Representatives, Committee on Science and Technology, Task Force on Science Policy. 1985. *Science Policy Study—Hearings Volume 8: Science in the Political Process,* 99th Congress, First Session (Washington, D.C.: U.S. Government Printing Office).

U.S. House of Representatives. 1988. *Omnibus Trade and Competitiveness Act of 1988: Conference Report to Accompany H.R. 3* (Washington, D.C.: U.S. Government Printing Office).

U.S. House of Representatives. 1980. *Toward the Endless Frontier: History of the Committee on Science and Technology, 1959–79* (Washington, D.C.: U.S. Government Printing Office).

U.S. House of Representatives. 1949. *Hearings before a Subcommittee of the Committee on Interstate and Foreign Commerce on H.R. 12, S. 247, and H.R. 359,* 81st

Congress, 1st Session, March 31, April 1, 4, 5, 26 (Washington, D.C.: U.S. Government Printing Office).

U.S. House of Representatives. 1948. *Hearings before the Committee on Interstate and Foreign Commerce on H.R. 6007 and S. 2385,* 80th Congress, Second Session, June 1 (Washington, D.C.: U.S. Government Printing Office).

U.S. House of Representatives. 1947a. *Hearings before the Committee on Interstate and Foreign Commerce on H.R. 942, H.R. 1815, H.R. 1830, H.R. 1834, and H.R. 2027,* 80th Congress, First Session, March 6 and 7 (Washington, D.C.: U.S. Government Printing Office).

U.S. House of Representatives. 1947b. *House of Representatives Report Number 1020,* 80th Congress, 1st Session (Washington, D.C.: U.S. Government Printing Office).

U.S. House of Representatives. 1946. *Hearings before a Subcommittee of the Committee on Interstate and Foreign Commerce on H.R. 6448,* 79th Congress, Second Session, May 28 and 29 (Washington, D.C.: U.S. Government Printing Office).

U.S. Senate. 1990. *Hearings before the Committee on Governmental Affairs on S. 1978,* 101st Congress, Second Session (Washington, D.C.: U.S. Government Printing Office).

U.S. Senate. 1948. *Senate Report Number 1151* (National Science Foundation), 80th Congress, 2nd Session (Washington, D.C.: U.S. Government Printing Office).

U.S. Senate. 1945/6. *Hearings before a Subcommittee of the Committee on Military Affairs Pursuant to S. Res. 107 (78th Congress) and S. Res. 146 (79th Congress),* 79th Congress, 1st and 2nd Sessions, October 8–12, October 15–19, October 22–26, October 29–31, November 1–2, 1945, March 5, 1946, Parts 1–6 (Washington, D.C.: U.S. Government Printing Office).

U.S. Senate, 1945a. *The Government's Wartime Research and Development, 1940–44.* Report from the Subcommittee on War Mobilization to the Committee on Military Affairs. Part II. Findings and Recommendations. July (Washington, D.C.: U.S. Government Printing Office).

U.S. Senate. 1945b. *Legislative Proposals for the Promotion of Science: The Texts of Five Bills and Excerpts from Reports* (Washington, D.C.: U.S. Government Printing Office).

U.S. Senate. 1943/4. *Hearings before a Subcommittee of the Committee on Military Affairs Pursuant to S. Res. 107 and on S. 702,* 78th Congress, 1st and 2nd Sessions, March 30, April 3, 9, 27, June 4, 17, October 14, 15, 21, November 4, 8, 24, December 9, 1943, February 8, 10–12, May 19, August 17, August 29, September 7, 8, 12, 13, 1944, Parts 7–16 (Washington, D.C.: U.S. Government Printing Office).

U.S. Senate. 1942. *Hearings before a Subcommittee of the Committee on Military*

Affairs Pursuant on S. 2721, 77th Congress, 2nd Session, October 13, 21, 22, 27, November 17, 18, 20, 25, 27, December 4, 10, 11, 12, 14, 17, 18, 19 Volumes 1–3 (Washington, D.C.: U.S. Government Printing Office).

U.S. Senate 1943. "The Mobilization of Science" (S. 702), *Science,* 97 (2523): 407–12.

Varmus, Harold E. and Marc W. Kirschner. 1992. "Don't Undermine Basic Research," *New York Times,* September 29, p. A15.

Vig, Norman J. 1975. "Policies for Science and Technology in Great Britain: Postwar Development and Reassessment." In T. Dixon Long and Christopher Wright (eds.), *Science Policies of Industrial Nations: Case Studies of the United States, Soviet Union, United Kingdom, France, Japan, and Sweden* (New York: Praeger), 59–109.

Waterman, Alan. 1960. "The National Science Foundation: A Ten Year Resume," *Science* 131 (May 6): 1341–54.

Waterman, Alan. 1951a. "The National Science Foundation Program," *Science* 114 (August 31): 251–52.

Waterman, Alan. 1951b. "Present Role of NSF," *Bulletin of Atomic Scientists* 7 (June): 165–67.

Waterman, Alan. 1949. "Government Support of Science," *Science* 110 (December 30): 701–7.

Weart, Spencer R. 1979. "The Physics Business in America, 1919–1940: A Statistical Reconnaissance." In Nathan Reingold (ed.), *The Sciences in the American Context: New Perspectives* (Washington, D.C.: Smithsonian Institution Press), 295–358.

Weinstein, James. 1968. *The Corporate Ideal in the Liberal State, 1900–1918* (Boston: Beacon Press).

Weir, Margaret. 1988. "The Federal Government and Unemployment: The Frustration of Policy Innovation from the New Deal to the Great Society." In Margaret Weir, Ann Orloff, and Theda Skocpol (eds.), *The Politics of Social Policy in the United States* (Princeton, NJ: Princeton University Press), 149–90.

Weir, Margaret, Ann Shola Orloff, and Theda Skocpol. 1988. "Understanding American Social Politics." In Margaret Weir, Ann Orloff, and Theda Skocpol (eds.), *The Politics of Social Policy in the United States* (Princeton, NJ: Princeton University Press), 3–35.

Weir, Margaret and Theda Skocpol. 1985. "State Structures and the Possibilities for 'Keynesian' Responses to the Great Depression in Sweden, Britain, and the United States." In Peter Evans, Dietrich Rueschemeyer, and Theda Skocpol (eds.), *Bringing the State Back In* (New York: Cambridge University Press), 107–63.

Whalley, Peter. 1986. "Markets, Managers, and Technical Autonomy," *Theory and Society* 15: 223–47.

White, Robert M. 1989. "Toward a U.S. Technology Policy," *Issues in Science and Technology* (spring): 35–36.

Whitley, Richard. 1984. *The Intellectual and Social Organization of the Sciences* (Oxford, England: Clarendon).

Wilensky, Harold L. and Lowell Turner. 1987. *Democratic Corporatism and Policy Linkages* (Berkeley: Institute of International Studies, University of California, Berkeley).

Wise, George. 1985. "Science and Technology," *Osiris,* 2nd series, 1: 229–46.

Wise, George. 1980. "A New Role for Professional Scientists in Industry: Industrial Research at General Electric, 1900–1916," *Technology and Culture* 21 (July): 408–29.

Wolfe, Dael. 1957. "NSF: The First Six Years," *Science* 126 (August 23): 335–43.

Wolfe, Dael. 1947. "The Inter-Society Committee for a NSF: Report for 1947," *Science* 106 (December 5): 529–33.

Wright, Erik Olin. 1985. *Classes* (London: Verso).

Wright, Erik Olin. 1978. *Class, Crisis and the State* (London: Verso).

Wynne, Brian. 1992. "Representing Policy Constructions and Interests in SSK," *Social Studies of Science* 22: 575–80.

Yamauchi, Ichizo. 1986. "Long Range Strategic Planning in Japanese R&D." In Christopher Freeman (ed.), *Design, Innovation and Long Cycles in Economic Development* (New York: St. Martin's), 169–85.

York, Herbert F. and G. Allen Greb. 1977. "Military Research and Development: A Postwar History," *Bulletin of Atomic Scientists* (January): 13–26.

Young, James H. 1980. "Merck, George Wilhelm." In John A. Garraty (ed.), *Dictionary of American Biography, Supplement Six, 1956–1960* (New York: Scribner's), 447–48.

Young, John A. 1988. "Technology and Competitiveness: A Key to the Economic Future of the United States," *Science* 241 (July 15): 313–15.

INDEX

Daniel Lee Kleinman is Assistant Professor in the School of History, Technology, and Society at Georgia Institute of Technology.

Library of Congress Cataloging-in-Publication Data

Kleinman, Daniel Lee.

Politics on the endless frontier : postwar research policy
in the United States / Daniel Lee Kleinman.

p. cm.

Includes bibliographical references (p.) and index.

ISBN 0-8223-1583-1 (cloth). — ISBN 0-8223-1598-X (pbk.)

1. Research—Government policy—United States—History.
2. Science and state—United States—History. 3. Science—
Social aspects—United States—History. 4. National Science
Foundation (U.S.)—History. I. Title.

Q180.U5K5 1995

338.97306—dc20 94-41440 CIP